Brol

MW01537707

The Theoretical Errors and Experimental Failures of the Standard Model of Particle Physics

E. Comay

This is a book on physics. The source of known issues is pointed out. New claims are adequately proved and presently accepted incorrect theories are disproved.

Copyright © 2021 by Eliahu Comay

First ebook edition November 2021

Cover Designer: nskvsky

ISBN 978-1-64153-417-8 (paperback)

Published by Eliahu Comay
eli@charactell.com

Foreward by Ofer Comay

This book deals with several problems in contemporary physics, but here, I want to address a specific topic—the theories that describe the internal structure of protons and neutrons.

With the inception of nuclear physics almost a century ago, physicists studied the forces acting between nucleons (protons and neutrons) that hold them inside the atomic nucleus. Even in this early phase, physicists noticed a resemblance between the van der Waals force holding noble atoms inside a drop of liquid and the nuclear forces holding nucleons inside the atomic nucleus. This resemblance is the basis for the Nuclear Liquid Drop Model. This model has been successful both quantitatively and qualitatively; it has been used for explaining alpha radiation (1928), incorporated into the Bethe–Weizsacker mass formula that explains the mass of the nuclei except for very light nuclei (1935), employed as a simple formula for the nuclear radius excluding very light nuclei (1935), and used to explain the fission of uranium 235 nucleus (1939). There are other effects that emphasize the similarity between the nucleus and a drop of liquid, such as the EMC effect (1983).

The success of the liquid drop model suggests that the behavior of nucleons in an atomic nucleus is similar to that of noble atoms in a liquid drop. However, the similarity between atoms and nucleons and between electrons and quarks is reflected in many other phenomena, most of which are unexplained by the dominant theory today. For example, the impressive resemblance of the potential graph between noble atoms and the potential graph between nucleons, the strong interaction of energetic photons with quarks that represents the sister phenomenon of the interaction of photons with electrons, and the parity and flavor conservation of both strong and electromagnetic interactions all have yet to be explained.

The immense resemblance is supposed to immediately raise the following issue: As for noble atoms that are in a liquid drop, we know what dictates the behavior of the forces acting between them: It is the structure of the atom that has a positive nucleus that attracts electrons arranged in shells. This is explained by quantum theory and electromagnetic formulas showing that the potential decreases inversely with the distance. That is, if we assume that nucleons are similarly constructed as well, where the nucleon has a core that attracts quarks which repel each other and the quarks are arranged in shells, we can almost automatically explain a broad set of phenomena in the domain of nuclear physics that remain unexplained to this day.

When I introduced the subject to an expert professor in the field, he immediately replied: "The idea has long been tested and completely ruled out." However, in reality, the idea has never been tested, and no scientific arguments have ever been presented to explain why it was rejected. The professor rightly assumed that this idea was so self-evident that it must have been tested from the outset, and since it is not an idea that dominates academia today, it has probably been completely ruled out. I suppose that this was the way of thinking that led to the retrieval of a wrong automatic response.

The accepted theory today that describes the structure of nucleons is called quantum chromodynamics (QCD). When QCD was invented, about fifty years ago, little was known compared with what we know today about this field. Still, the theory was constructed as building a new game – a collection of laws with almost no limit to the freedom given by the inventors of the theory to create new physics. These are as follows: They invented three kinds of charges, called colors; arbitrary laws that do not allow particles with an unequal number of colors to exist independently; force laws that work the opposite of any other force, and by which the particles pull each other stronger as they move away from each other; and massless particles, gluons, that exist in the nucleons and cannot come out. The "highlight" of the theory is an equation that is unsolvable using ordinary tools. (Supercomputers work hard to find approximate solutions for this equation.) This weak point of the theory has become its main strength: It is difficult to refute a theory if we have not yet been able to accurately solve its central equation fifty years after its invention.

Anyone familiar with QCD should suspect that the invention of QCD is an academic "exercise" in the philosophy of science. It is an exercise that aims to show that, for each set of physi-

v

cal phenomena—with the help of fertile imagination, mathematical abilities, and unlimited freedom in inventing new laws of physics— we can build a theory that will explain these phenomena. And because of this, it is clear that there is no connection between such a theory and reality because we had many ways to build such a theory and we chose only one of them. But oddly enough, QCD has taken a firm grip among particle physicists, and despite a long series of experimental refutations, it is still the leading theory without competition.

It is amazing and disappointing to see how serious researchers try to explain phenomena that seem to contradict QCD without having the courage to challenge QCD itself. Here is a striking example: According to QCD, there are no massive particles within nucleons other than the three quarks (and quark-antiquark pairs, as must be present according to quantum field theory). Physicists were stunned when it was discovered in the 1970s at SLAC that much of the proton's momentum was not carried by the quarks. But even then, only a year or two after the publication of QCD, they feared to challenge the theory and argued that the gluons are the ones bearing the missing mass. In the last two decades, with the increase of energies in particle accelerators, more data have been collected. The increase in the proton-proton cross-section and the proportion of the elastic proton collisions (at high energies, nearly 30% of proton collisions are elastic collisions) was discovered. There is only one explanation for the new data: There is a massive rigid core inside the nucleons in addition to the three valence quarks. Thus, physicists finally started to talk about new massive particles inside the nucleons. This refutes the basis of QCD. But even now, scientists are trying to explain the same massive particles in a proton, the "pomerons," as part of QCD, although QCD was invented in the first place with the assumption that there are no massive particles in nucleons other than the three quarks.

The theory explaining the nuclear liquid drop model and many hadronic phenomena was published by my father, Eliyahu Comay, the author of this book, in a series of publications beginning in 1984. The first publication developed a regular theory of magnetic monopole. The results obtained describe the equations of the quarks. According to these equations, the quarks carry a negative strong charge that repels a strong charge of the same type, just as electrons carry a negative electric charge that repels negative electric charges. From those equations, it turns out that the strong charge does not produce a direct interaction with an electric charge, but from the development of the equations, it follows that its radiation particle is the photon, exactly the same known photon that interacts with an electric charge. For this reason, the strong charge in the article (and in this book) is called a magnetic monopole, although it is not the same magnetic monopole of Dirac, and although it does not produce a direct interaction with an electric charge as an active magnetic dipole is supposed to produce.

My father's theory was based on the variational principle. The importance and centrality of the variational principle are accepted today by all physicists. That this monopole derives directly from the variational principle constitutes tremendous support for this theory.

By the way, Dirac suspected that the proton contains magnetic monopoles, but according to the formulas he developed, there was supposed to be a direct interaction between electric charges and Dirac monopoles, which does not happen in reality. Seemingly for this reason, the idea was abandoned.

The new monopole theory has been published in parts in many articles, and a concluding article was published in 2012. This article led to its exposure to a wide audience. But surprisingly, there was not a single scientist who dared to lift the glove and try to disprove or support the theory. In fact, the opposite happened: Immediately after the publication, my father was completely denied the ability to publish his articles on the arXiv website, which is generally available to researchers and allows them to publish their articles before they are sent to peer-reviewed journals. That is, instead of facilitating a free scientific discussion in academia, unknown forces acted to make it even more difficult to discuss the subject. And to this day, institutional scientists make no reference to the possibility that the theory they hold is incorrect, and not a single article has been published other than by my father regarding his theory. I wonder when scientists will begin to seriously discuss this important issue that describes what the smallest particles of matter consist of and openly consider the validity of the dominant theory, QCD. My guess is never! I would love to be proven wrong.

Ofer Comay

October, 2021

Contents

Preface

I've seen practical advice given to ordinary physicists:

Rule #1: "Shut up and calculate!"

Some people have probably adopted this advice. However, one result of this state of affairs is that any attempt at error correction of any theoretical element of contemporary physics is strongly rejected by most members of the particle physics establishment. Personally, I've been unaware of this advice, and I've taken only its second part. During my scientific life, I've stumbled on too many people who have apparently followed it. This book is one result of my attitude concerning Rule #1.

A not negligible portion of the work that has produced this book has been dedicated to the task of error correction – both in physics and in this book. However, it is well known that human work is not always error-free. I thank in advance everybody who will mention erroneous points of this book. They may be correctable errors, like misprints, or even fundamental errors in theoretical physics.

Error correction is a common element of a team of persons trying to accomplish a certain assignment. Factories call it QA, and computer programmers call it debugging. I'm quite sure that the person in the street does not doubt that scientists in general and physicists in particular, follow suit. Unfortunately, as stated above, my broad experience has taught me that this practice does not hold in the domain controlled by particle physicists. On the contrary, any attempt to discuss an apparently erroneous point is automatically rejected, and the person who dares to do so is treated as an enemy.

An example that is taken from my experience illustrate this issue. It briefly describes the changes of my status in the arXiv, which is a vehicle for a prepublication of papers that discuss physical problems. At arXiv's inception, I had the right to publish in it. Furthermore, I've been an endorser with the right to recommend publications of papers that have been written by persons who still do not have this right. (Here is a link to an article that I posted on arXiv in 2004: `https://arxiv.org/pdf/physics/0405050.pdf`.) However, I've found that somebody has canceled my right to publish in most of the arXiv's sectors. At present, I'm allowed to publish only in the *General Physics* sector. A colleague has told me that this sector has the lowest reputation... Personally, I do not care much about reputation. This is not the end of the story. Indeed, I've realized that for the last 8 years, some people called "moderators" have systematically removed every paper that I had posted on arXiv. (In my view, an Orwellian organization is likely to label people who brutally reject papers by the term *moderators*.) The plain meaning of this activity of arXiv's managers is that they prevent some people who use arXiv from being aware of my work.

Another example is a quotation from a review report on Woit's book – *Not Even Wrong* – which criticized String Theory. A qualified theoretical physicist says: (see `https://www.amazon.com/review/R1ZJZAZXO7G9QA`). "But the string community (or any other scientific group) cannot be allowed to turn their chosen approach into a fad, a cult, a religion or - worst - an inquisition. That might be sociologically amusing if it weren't so pernicious for physics and physicists. The effect is that physics departments become the monopoly of self-perpetuating, self-congratulating clans of homogenous thinkers. (Lee Smolin uses the psychological term 'groupthink'.) Young physicists are informed that string theory 'is the only game in town'. Sign up for it or go away."

My conclusion is that too many people who run the particle physics establishment automatically reject every examination of the possibility that an erroneous element exists in the presently accepted theories. This is an unfortunate situation because a new physical idea that relies on an erroneous basis is doomed to add further incorrect complications to an already erroneous structure.

The foregoing lines indicate an important sociological problem with how scientific work is organized today. This is indeed a crucial problem that is discussed in many places. Let me mention just a few publications here. The book *Against the Tide, A Critical Review by Scientists of How Physics and Astronomy Get Done*, edited by Corredoira and Perelman contains several articles that discuss this issue. The books and publications of P. Woit and L. Smolin show how string theory proponents unjustifiably dominate the present particle physics research.

I deeply support every endeavor aiming to promote free discussion of scientific problems. However, this book does not discuss any sociological aspect of contemporary organizations of physicists. Rather, it is dedicated to a critical examination of the physical elements of the *Standard Model of Particle Physics*. The long list of physical errors of this model that are proved in this book *indirectly* show sociological problems of the community that adheres to this false assembly of theories.

This book is not a comprehensive textbook that discusses every relevant detail. Its contents can be divided into the four following sets:

C.1 Well-known topics that can be found in standard textbooks. (Yes, contemporary textbooks include some correct theoretical elements!) SR and the Lagrangian density play a primary role in the structure of a quantum theory of an elementary particle. The outcomes of these theoretical elements are pointed out together with other issues.

C.2 Constraints on a physical theory of an elementary particle play a crucial role in this book. They are listed in section 3.9 and summarized near the end. A knowledgeable reader can observe that they are all taken from textbooks. In contrast, this book proves that supporters of the Standard Model of particle physics (SM) do not respect some of these constraints.

C.3 New theoretical relations that I originally found. Most of them have already been published in my papers. However, some new issues are published in this book for the first time.

C.4 Erroneous elements of the SM. Corrections of some of these topics are shown. In my opinion, a discussion where one party plays the role of devil's advocate can only improve the state of theoretical physics.

I do not know what to say about requirements that enable a reader to understand the book. Evidently, the higher the physical education, the better. However, this book contains many elements that are not discussed in the mainstream literature. Hence, even if readers understand 30%-40% of the topics discussed in the book, they will acquire a lot of information that cannot be found in mainstream textbooks at present. For example, several parts of this book discuss problems of electrodynamics and general physics (chapters 1-8). (Subsection 4.3.2 *proves* the peaceful coexistence of the concepts of point-like particle and wave.) These parts should be understandable by everyone who has finished graduate studies in physics, as well as some mathematicians, chemists, and engineers.

Proofs of many SM errors are discussed in the book. In my opinion, there are two explanations for why this is an important part of the text: Erroneous elements prevent a correct development of physics, and mainstream physicists strongly reject any attempt to discuss SM errors. Therefore, this book departs from the standard practice of brief presentation of topics, and some SM errors are discusses more than once.

The contents of about the two pages of section 9.2 on p. 114 can be easily read. I'm quite sure that this short text will convince readers that this book relies on solid ground. Similarly, table 14.1 on p. 220 together with the accompanying text, provide easily understandable arguments that support the book.

There is another aspect of the notion of understanding a physical topic. Here I wish to explain this point by means of examples. Undergraduate students that study electrodynamics learn Maxwell equations. These are partial differential equations of the electromagnetic fields. In many cases, the teacher presents several examples of problems and solves them. Students are requested to solve other problems at home. Here students get the feeling that they really understand the meaning of these equations. Students also learn about the Lorentz force, and how electromagnetic fields exert force on a charge. Thus, they come to feel that they understand the interrelations between mechanics and electrodynamics.

However, things are not perfect, and in some cases, time is short. Thus, every once in a while, students realize that they must learn things by heart, without having a comprehensive understanding of the coherent interrelations between all relevant topics. This is certainly not the best situation, but I'm quite sure that it happens to many students. (I've found myself in this situation quite a few times...) I wish to show an example that proves my claim.

Several generations of particle physics students have already studied electroweak theory, representing the weak interactions SM sector. Electroweak theory is a quantum field theory. Students who learn about it have studied Maxwellian electrodynamics and the Dirac theory of the electron, which are also field theories. In the latter cases, students have seen the explicit form of the partial differential equations of electromagnetic fields and the Dirac field. Furthermore, they have probably solved these equations for several (not too complicated) physical systems. These solutions have enabled them to see a wider picture of physics and the interrelations between theory and experiments.

About 50 years have elapsed since the birth of the electroweak theory. However, textbooks on this theory still do not show the *explicit form* of the differential equations of this theory. A fortiori, no solution to these unknown equations is compared with experimental results.

One may derive several conclusions from the facts mentioned above:

C.1 Particle physics students learn electroweak theory by heart.

C.2 To understand the general meaning of the previous argument, one does not need to know specific details of partial differential equations. It is enough to know that these equations are an important part of a field theory and that electroweak theory still does not have these equations.

C.3 A reader with some general scientific education who understands the previous argument can understand some parts of this book.

The contents of many subjects of this book are included in articles that I have published in scientific journals. Many items of chapter 10 are included in an article that was published in the Electronic Journal of Theoretical Physics, **9**, 93-118 (2012), `http://www.ejtp.com/articles/ejtpv9i26p93.pdf`.
The subjects are organized in this book in a compact form that facilitates the reading of many parts of my scientific work.

Readers generally begin reading a book from its beginning. This is okay. However, this book may be regarded as a "dissident" book, and some readers may take a skeptical approach to its contents. I advise these readers to begin reading the Epilogue chapter on p. 241. I think that this chapter will convince such readers that the issues discussed in this book deserve a close examination. Analogously, as stated above, the contents of about three pages at the beginning of section 14.1 on p. 220 are easily read. I'm quite sure that this short text will convince readers that this book's arguments rest on solid ground.

Knowledgeable readers may start reading this book at the beginning of its Conclusions chapter on p. 225. That chapter begins with a description of the physical principles that this book adheres to. In short, it shows that this book seriously adheres to fundamental elements of physics.

Here I wish to make a minor remark. Unlike most journals, the references that are mentioned in this book do not take a well-defined form. For me, an appropriate description of the published object is enough.

Many people have helped me in one way or another to accomplish some tasks of my scientific work. In particular, I wish to thank my family members for their encouragement and support. Other people have also helped me. Prof. S. Rosset, Prof. Z. Schuss, and Prof. D. Levin from the School of Mathematics, Tel-Aviv University, Prof. K. T. Hecht and Prof. J. Janecke from the University of Michigan, Ann Arbor, K. Hellreich, who lived in Ann Arbor in 1983, A. Ney from Paris, C. Botner, P. Einat, Y. Tal, Y. Lev, and M. Meiri from Israel. I ought special thanks to N. Zeldes and I. Kelson, my M.Sc. and Ph.D. instructors, who have taught me specific physical topics and general issues on how to carry out scientific research. Correspondence with Dr. G. Bella of the School of Physics, Tel Aviv University is acknowledged.

Eliahu Comay

October, 2021

Chapter 1

Introduction

This book adheres to the principle that Wigner has described as: "The Unreasonable Effectiveness of Mathematics in the Natural Sciences" [1]. Here is a historical example of the benefits that mathematics brings to physics. The mathematician A. M. Legendre lived about 200 years ago. One of his mathematical items is the Polynomials of Legendre. A century later his polynomials have been used for proving that quantum mechanics adequately describes the angular part of the electronic states of the hydrogen atom. This evidence is one of the cornerstones of quantum mechanics.

The relationships between mathematics and physics have many aspects. Let us begin with an amusing tale that illustrates how a mathematician finds a shorter logical path for solving a problem. Students in general and students of close disciplines, in particular, try not to miss an opportunity to poke fun at other groups of students. Here is an example. A physicist and a mathematician are asked to solve two "problems".

P.1 You are in an apartment and its kitchen comprises a gas cooker, a kettle, and a water tap . There is a matchbox in the living room.
Problem: How do you boil water?

The physicist's solution: I take the kettle, open the tap and put an adequate amount of water in the kettle. Then, I put the kettle on the gas cooker, go to the living room, take the matchbox, go back to the kitchen, light the gas, and wait until the water boils.

The mathematician's solution is the same.

P.2 This is a similar problem, but the matchbox is in the kitchen.

The physicist's solution contains an appropriate repetition of his quite lengthy solution to the first problem.

The mathematician's solution goes as follows: I take the matchbox and put it in the living room. I've already solved this problem.

Mathematics provides powerful tools that enable theoretical physicists to do a good job in much more profound ways than this tale suggests. This book is a modest attempt to describe this issue. It concentrates on the following relationship between mathematics and theoretical physics:

> *A physical theory must have a coherent mathematical structure.*

This book applies three kinds of requirements:

Req.1 Any physical theory should have a coherent mathematical structure.

Req.2 The mathematical structure of any given theory should agree with well-established physical principles. These principles take a mathematical form, but they are based on solid experimental measurements.

Req.3 In addition to these requirements, a physical theory must satisfy another kind of constraint: it must adequately describe well-established experimental data that belong to its domain of validity.

This book argues that these criteria are highly important elements of physics. Thus, the identification of an erroneous structure of a given theory justifies its rejection. In identifying such erroneous structures, the scientific community can save many fruitless efforts from being wasted on an inherently uncorrectable theory.

Taking a stringent mathematical approach for an examination of the coherence of a physical theory, one may think that a single genuine contradiction justifies the rejection of the examined theory. This is correct. However, this book refrains from this quite brief approach. The main reason for this is that people are likely to be convinced by many different proofs of the same assertion. Hence, this book shows quite a few different arguments that prove that the SM sectors – Quantum Electrodynamics (QED), Quantum Cromodynamics (QCD), and electroweak theories – are full of errors.

Most of the errors belong to Req.2 and Req.3, as pointed out above. The theoretical structure of the SM sectors violates physical principles. An analogous situation is found in the SM failures to comply with well-established experimental data. Many examples of these failures are discussed in this book. This means that mainstream experimental physicists work hard to provide the data, and the establishment sweeps their meaning under the carpet.

An example illustrates the problematic SM situation. Objections to the SM have been published in many places. However, most of these objections have been published in books and journals that do not belong to the present physical establishment. An exception is a Letter that was published in Physics Today, a publication of the American Institute of Physics (see Physics Today **42**, 11, 13 (1989); doi: 10.1063/1.2811203). In this Letter, Martin Macháček compares the SM to the astronomical geocentric theory of Ptolemy. He claims that the SM is inherently wrong and uncorrectable. It turns out that after more than 30 years, no physicist has published an objection to Macháček's Letter.

This argument justifies the claim that SM supporters have no good answer to the claims of Macháček's Letter. Indeed, suppose that instead of objecting to the meaning of the SM, Macháček argued against SR. Here, I would bet that many physicists would write papers that deny his claims and support SR. I conclude that SM supporters have no good answer to Macháček's challenge.

The general structure of this book is as follows: The beginning chapters present theoretical issues that are regarded as correct elements, with short descriptions of these topics. Further issues are derived from these elements. Subsequent chapters use these matters and prove that the SM is full of errors. Another aspect of this book is the derivation of correct theoretical elements that belong to the topics that are discussed here.

1.1 Notation and Units

This book uses several conventions: A system of units where $\hbar = c = 1$ is used. A power of length $[L^n]$ denotes the dimension of every term. The Lorentz metric is diagonal and its entries are (1,-1,-1,-1). Greek indices run from 0 to 3, and (in most cases) Latin indices run from 1 to 3. Standard relativistic notation is used. $\epsilon^{\alpha\beta\gamma\delta}$ is the completely antisymmetric unit tensor of the fourth rank, and $\epsilon^{0123} = +1$. The Dirac $\boldsymbol{\alpha}$, β, and γ^μ matrices take the form that is used in [2].

The variational principle plays a primary role in this book. A Lagrangian is used in the case of classical particles, and a Lagrangian density is used for electromagnetic fields and quantum particles. Hence, the celebrated textbook on classical electrodynamics [3] uses hybrid expressions – a Lagrangian describes the motion of point-like charged particles, and a Lagrangian density describes the laws of electromagnetic fields. Here, the term Lagrangian density is sometimes used for classical particles.

Magnetic monopoles take an important part of this book. In electrodynamics, the word charge denotes electric charge. Therefore, the words "magnetic charge" may confuse readers. Following Dirac, the words "magnetic strength" or "monopole strength" denote the magnetic analog of the electric charge [4]. Please note that the word "strength" is also used in other cases.

1.2   Acronyms

The following acronyms and terms are used in the book. However, the explicit unabridged words are used in some cases.

1. AB – Ahronov-Bohm

2. BE – Binding energy

3. CDM – Cold dark matter

4. CI – Configuration interaction

5. CP – The charge conjugation and parity symmetries

6. CPH – Classical Physics

7. DAEI – Dual attribute of electromagnetic interactions (see subsection 11.6.3)

8. DDWIT – Dipole-Dipole Weak Interaction Theory (see section 11.3)

9. The dog-tail constraint – (see section 3.2)

10. EM – Electromagnetic

11. EMC – European muon collaboration

12. GCP – Generalized Correspondence Principle

13. GUT – Grand Unified Theory

14. KG – Klein-Gordon

15. LHCD – Lagrangian-Hamiltonian Covariance Difference (see section 7.4)

16. MLE – Maxwell Lorentz Electrodynamics (see subsection 3.6.1)

17. Monopole – Magnetic monopole

18. QCD – Quantum Chromodynamics

19. QED – Quantum Electrodynamics (as known in 2021)

20. QFT – Quantum Field Theory

21. QM – Quantum Mechanics

22. RCMT – The Regular Charge-Monopole Theory

23. RQM – Relativistic Quantum Mechanics

24. SM – The Standard Model of particle physics

25. SQM – Strange quark matter

26. SR – Special Relativity

27. SUSY – Supersymmetry

28. VE – Variational Electrodynamics (see subsection 3.6.1)

29. WIMP – Weakly interacting massive particle

Chapter 2

General Topics

This chapter presents two general topics: the meaning of theories and models and the Occam's razor criterion.

2.1 Theories and Models

An important element of science is establishing a clear definition of concepts using specific words. For example, in the realm of mathematics, it is stated: "There is a reason for special notation and technical vocabulary: mathematics requires more precision than everyday speech. Mathematicians refer to this precision of language and logic as 'rigor'" [5]. Theoretical physics uses mathematical structures. Therefore, it has the same requirements to bring about coherence in the discipline. This section is dedicated to the notions of theory and model in theoretical physics.

A review of the literature indicates that the distinction between a theory and a model is still obscure. For example, philosophers of science still have no unique definition for the meaning of a model [6]. Another example of this unclear situation is the SM. Wikipedia's policy is to adhere to the consensus. At present (August 2021), its SM item begins with this statement: "The Standard Model of particle physics is the theory describing three of the four known fundamental forces (the electromagnetic, weak, and strong interactions, while omitting gravity) in the universe, as well as classifying all known elementary particles" [7]. This is an unfortunate situation because, as stated above, a clear meaning of words is required for scientific work. For this reason, this book adopts the following definitions of theory and model, which rely on practical issues:

1. A physical theory satisfies the three requirements mentioned
   in the Introduction section. Because of their importance and
   brevity (under 10 lines), they are repeated here:

   Req.1 A physical theory should take a coherent mathematical
         structure.

   Req.2 The mathematical structure of any given theory should
         agree with well-established physical principles. These
         principles take a mathematical form, but they are based
         on solid experimental measurements.

   Req.3 In addition to these requirements, a physical theory
         must satisfy another kind of constraint: It must ade-
         quately describe well-established experimental data that
         belong to its domain of validity.

   A theory is unacceptable if it fails to abide by any of these requirements. A practical examination of a good theory shows
   that it has agreeable elements like relatively high accuracy
   of its predictions, successful predictions of a large variety of
   different experiments, and no failure in cases that belong to
   its domain of validity.

2. A model is not expected to satisfy the stringent requirements
   imposed on an acceptable theory. It is used in cases where
   there is no good theory or in cases where a relevant theory
   yields infeasible mathematical expressions.

   Nuclear physics uses quite a few models. For example, the
   nuclear shell model (see, e.g., [8], pp. 250-271; [9]) explains
   the J^π of nuclear states, the magic numbers, the relations
   between nuclear energy levels, and so on. The *nuclear liquid
   drop model* aims to describe other nuclear properties, such
   as nuclear mass, nuclear radius, and processes like fission [8].
   The Garvey-Kelson mass relations represent another example of a nuclear model. Here, the difference between the sum
   of the masses of three nuclei and the sum of the masses of
   three other nuclei is quite small. Generally, it takes a 2-
   digit number of KeV. These relations enable a nuclear mass
   calculation. As in the case of a typical model, these calculations yield good results for interpolation, whereas their
   quality deteriorates with extrapolation (see, e.g., [10–12]).
   Solid state physics examines systems that comprise an astronomical number of electrons and nuclei. Hence, ordinary
   electron theories, such as those that yield the Schroedinger,

Pauli, and Dirac equations, are infeasible for the calculation of crystals and so on. Therefore, models like the *nearly free electron model* are used [7]. All these models yield predictions where the accuracy falls far below the accuracy of results of an ordinary physical theory.

These points explain the inherent differences between a theory and a model. A theory may be discarded because it is wrong, whereas the usage of a model depends on its practical applicability for a narrow group of cases. For example, even if the original nuclear liquid drop model does not explain the nuclear magic numbers, it is still not discarded. On the contrary! It may be improved by an addition of new terms aiming to account for these numbers.

> Conclusion: A physical theory is examined by rigorous mathematical criteria to determine whether it is true or false. A physical model is examined by practical elements like its usefulness for a narrow set of physical effects.

This book adopts the general approach which regards the SM as a *set of theories*. A special attention is dedicated to the validity of theories that apply to strong, electromagnetic, and weak interactions.

2.2 The Occam's Razor Criterion

The Occam's razor criterion refers to a case where one wants to choose between two different interpretations (i.e., theories) that explain the same set of effects. It says that the theory that depends on a smaller number of assumptions is the right theory [13].

Occam's razor is not regarded as a rigorous argument that is a sufficient reason for a theory disqualification; while this book generally agrees with that assertion, the criterion can be regarded as a rule of thumb or a supporting argument in the case of choosing between two theories. This book mentions the Occam's razor criterion in this sense.

Another aspect of the Occam's razor criterion is the difference between the number of assumptions of the competing theories. For example, if one theory relies on six assumptions and the other relies on seven, then the Occam's razor criterion looks useless. In contrast, if one theory relies on six assumptions and the other relies on more than thirty assumptions, then the Occam's razor criterion is decisive!

Chapter 3

Cornerstones of Particle Theories

This chapter describes elements of theoretical physics that are used in this book. Some of these elements are taken for granted, whereas supporting arguments are added for others.

PR.1 Special relativity.

PR.2 Maxwell equations of the electromagnetic fields.

PR.3 The correspondence between theories that are connected by hierarchical order, like Newtonian mechanics and relativistic mechanics, where the low-velocity limit of the latter agrees with the former.

PR.4 The de Broglie principle.

PR.5 The variational principle that relies on a Lagrangian density of a given quantum particle. Its general form is

$$\mathcal{L}(\psi, \psi_{,\mu}),$$

where $\psi(t, \boldsymbol{x})$ is the generalized coordinate of the Lagrangian density. Every term of this Lagrangian density is a Lorentz scalar with a dimension of $[L^{-4}]$. Its interaction with the external field ϕ takes this form

$$\mathcal{L}(\psi, \psi_{,\mu}, \phi, \phi_{,\mu}),$$

A Lorentz scalar Lagrangian density ensures that the theory abides by SR (see [14], p. 35).

PR.6 The Noether theorem.

PR.7 A quantum theory must have a coherent equation of motion derived from the variational principle.

PR.8 A quantum theory must abide by Wigner's analysis of the unitary representations of the inhomogeneous Lorentz group.

PR.9 A quantum theory must explain experimental data that belong to its validity domain. This objective requires that every Lagrangian density of a given quantum particle has an interaction term.

3.1   The Role of Mathematics in Theoretical Physics

A common premise that all physicists and engineers probably agree to says that the behavior of the physical world can be described by a mathematical language. Wigner mentions many aspects of this issue in his well-known article [1]. He says:

> "Mathematical concepts turn up in entirely unexpected connections."

Three examples of amazing and entirely unexpected points that illustrate Wigner's statement are as follows:

- Charge conservation may be regarded as a reasonable element that describes physical processes. Physicists agree on this matter, and probably, the person in the street will also agree on it. Maxwell has used this point and constructed his famous equations – namely, partial differential equations that are the laws of the time evolution of electromagnetic fields. Solutions to these equations prove many things. Two unexpected examples of results of these equations – results of the electromagnetic laws that were still unknown in Maxwell's time – are as follows:

  - Light is a kind of electromagnetic wave.

  - The speed of light takes the same value in every inertial frame.

- Matrix algebra has definite laws. These laws enable a representation of a group by matrices. The group SO(3) has three generators, and the commutation relations between these generators are:

$$[L_x, L_y] = L_z, \quad [L_y, L_z] = L_x, \quad [L_z, L_x] = L_y. \qquad (3.1)$$

  Another group is called SU(2). There are three Pauli matrices where the corresponding commutation relations are similar, and the difference is a factor of $2i$ (see [15], p. 268).

  Mathematicians may note this similarity and continue with their work. However, it turns out that this similarity enables the validity of a fundamental physical requirement – the correspondence between quantum mechanics (QM) and classical physics (CPH). In particular, the SU(2) group enables a quantum description of a spin-1/2 particle. The magnetic field of the electron's spin (a quantum object) is the same as the magnetic field of a current loop (a classical object). The identity of these fields requires the above-mentioned analogy between SO(3) and SU(2).

- Wigner's article [16] analyzes the unitary representations of the inhomogeneous Lorentz group. This is a mathematical work that was published in a mathematical journal. One of his results is as follows: A massive quantum state should have a definite spin J and $(2J + 1)$ components of J_z. A massless particle has helicity, where J_z takes only two values. This outcome has been confirmed in experiments. Thus, an atom with a total angular momentum J has $(2J + 1)$ spin projections. In contrast, the spin-1 photon has only two values: $j_z = \pm 1$.

A primary part of this book investigates the mathematical properties of physical theories. It turns out that experiments support coherent mathematical relationships and deny faulty ones.

3.2 Correspondence Between Physical Theories

Before entering a discussion about the merits of the generalized correspondence principle (GCP), one should define the acceptance criteria for a given physical theory. In this case, one should not go too far and argue that if a given theory contradicts certain experimental results then it should be automatically rejected. In such a case, one must examine the circumstances in which the experiment has been performed. For example, Newtonian mechanics provides a good description of experimental results in cases where the particle's speed is much smaller than the speed of light and the interparticle distance is long enough (or the angular momentum is much larger than $\hbar$). Thus, an experimental physicist may use Newtonian mechanics if the inaccuracy of this theory is smaller than the inaccuracy of the experimental device. An engineer uses Newtonian mechanics if the inaccuracy of this theory does not affect the operation of the designed apparatus. The validity domain notion of a given theory plays an important role in an examination of the theory's acceptability. The validity domain of a given theory is restricted to experimental conditions where results are properly explained by the theory. For example, the conditions mentioned above determine the validity domain of Newtonian mechanics. However, one should take very seriously Rohrlich's remark: "the validity limits are not drawn *ad hoc*. When a particular phenomenon is not accounted for by the theory, we cannot gerrymander the validity limits of the theory in order to "save the theory" (see pp. 1-6 of [17]).

The notion of validity domains can be used for defining hierarchical relationships between theories. This issue is briefly explained here. Consider two theories A, B, where theory B explains all the experimental results that are explained by theory A, and some other experimental results that are not explained by theory A (see Figure 3.1). This means that theory B has a higher rank in the hierarchical order of these theories. The relationships between Newtonian mechanics and relativistic mechanics illustrate this issue.

Evidently, theory B has a more profound physical meaning because it explains more kinds of experiments. Thus, the meaning of theory A may be considered "less significant." However, the merits of theory A should not be underestimated. This point is empha-

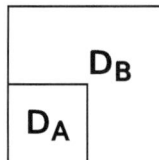

Figure 3.1: *The validity domain of theory A is included inside that of theory B.*

sized here. The validity domain of theory B contains cases where theory A works fine. Hence, in the validity domain of theory A, values of physical variables obtained from theory B must agree with the corresponding values obtained from theory A. It can be said that the limit of variables obtained from theory B must agree with the corresponding variables of theory A. *This means that the lower rank theory A imposes constraints on the acceptability of the higher rank theory B*. This relationship between theories A, B can be described as the tail wagging the dog. The expression "dog-tail constraints" is used below as a figurative description of this type of relationship between theories. For example, it is well known that for a mechanical system where the particles' velocity is much smaller than the speed of light, quantities obtained from relativistic calculations agree with the corresponding quantities of Newtonian mechanics. An analogous requirement has also been recognized by the founders of QM, and indeed, the Ehrenfest theorem proves the agreement of the classical limit of QM with classical physics [18].

Applications of dog-tail constraints can be found in many branches of physics. Einstein explains how Maxwellian electrodynamics yields expressions that agree with electrostatics in cases where the charges do not move (see [19], pp. 85-86). He also explains how General Relativity reduces to SR in the case of a weak gravitational field. In addition, Rohrlich points out multiple examples of physical theories that have hierarchical relationships (see pp. 1-6 of [17]). In so doing, he demonstrates the applicability of the GCP to the structure of physical theories. It is interesting to note that the relevance of the GCP to the relationship between quantum field theory (QFT) and QM is also stated clearly in Weinberg's well-known QFT textbook:

> *"First, some good news: quantum field theory is based on the same quantum mechanics that was invented by Schroedinger, Heisenberg, Pauli, Born, and others in 1925-26, and has been used ever since in atomic, molecular, nuclear and condensed matter physics"* (see [20], p. 49).

Remark: Because of historical precedence, the *Correspondence Principle* concept denotes the correspondence relationships be-

Figure 3.2: *Hierarchical relationships between 4 theories.*

tween QM and CPH. The more general version of this concept is called the *Generalized Correspondence Principle (GCP)*.

This book mainly uses the correspondence relationships between four theories (see fig. 3.2). Relativistic quantum mechanics (RQM) is inadequate for states like that of the proton where additional quark-antiquark pairs are explicitly found. Here, QFT is better. QM is inadequate for describing relativistic effects; here, RQM is better. CPH does not explain quantum effects (e.g., why the ground state of the hydrogen atom is stable and the electron's acceleration does not emit radiation); here, QM is better.

The GCP is not just an abstract theoretical structure. Examples of its applications are discussed in several places in this book. One case is given in the next subsection.

3.2.1 An Example

The following example illustrates how the dog-tail constraint works. Consider an inelastic collision of two particles that produces new particles. Astronomical numbers of such events have been recorded since the beginning of the accelerator era. The chronological order of this process is as follows:

1. First, two particles move in external electromagnetic fields. Relativistic classical mechanics and classical electrodynamics describe the state.

2. The two particles are very close to each other. RQM describes the state.

3. The two particles collide and interact. New particles are created. QFT describes the process.

4. Particle creation ends but they are still very close to one another. RQM describes the state.

5. All particles depart. Relativistic classical mechanics and classical electrodynamics describe the state.

CPH tells that energy and momentum are well-defined quantities in the initial and in the final stages of this process, and their final values abide by energy-momentum conservation. This means that the specific values of the total energy-momentum of the final state agree with the corresponding quantities of the initial state. Furthermore, the initial and the final states are connected by processes that are described by RQM and QFT. In particular, the

process of new particle creation is described only by QFT. Therefore, RQM and QFT must "tell" the final state the precise initial
values of the energy-momentum. It follows that a consistent form
of RQM and QFT must provide an adequate expression for the
Hamiltonian, and the limit of the expectation value of this Hamiltonian should agree with the energy of CPH. This simple example
shows how the dog-tail constraint works in the important case of
an inelastic scattering event.

3.2.2   Elements of the Classical Theory

Why the classical limit of quantum theories is an important theoretical element has been explained above. Details of several physical quantities that are relevant to a classical particle are discussed
below.

Landau and Lifshitz prove that an elementary relativistic particle is point-like (see [3], pp. 46, 47). The classical Lagrangian
takes advantage of this issue, and its form is

$$L(q, \dot{q}, t) \tag{3.2}$$

where q denotes a set of the particle's generalized coordinates (see
[21], p. 2). In particular, q of a free elementary classical particle
may be the ordinary Cartesian coordinates, where each point is
uniquely defined.

An alternative formulation of classical mechanics uses the
Hamiltonian

$$H(q, p, t) \tag{3.3}$$

where p denotes the generalized momentum $p_i = \partial L / \partial \dot{q}_i$ (see [21],
p. 131; [22], p. 337).

Solutions of the classical equations of motion that are derived
from (3.2) determine the particle's coordinates and its velocity as
functions of time. Alternatively, solutions of the canonical equations that are derived from (3.3) yield the particle's coordinates
and its momentum as functions of time.

An important attribute of the classical limit is that it is a
continuous process. For example, quantum theories describe the
time evolution of an electron that is bound to an atom. Hence, it
is known that solutions of the Schroedinger equation provide the
probability that the electron will be within a certain atomic volume. An ejection of the electron increases the atom-electron distance. The quantum theory that describes this process takes the
form of differential equations. Their solution indicates that this

is a time-dependent continuous process, and eventually, CPH explains the electron's behavior. This continuous process means that dimensions of a physical variable, which are not continuous variables, do not change in the transition between quantum theories and CPH. These issues affect the correspondence between quantum theories and CPH. Some specific topics are as follows:

CL.1 The solution of classical equations of motion of a point-like massive particle contains a well-defined value for its time-dependent position. Hence, quantum theories of a massive particle must define a variable where the classical limit reduces to a spatial point. The Schroedinger QM theory defines the particle's density, and its appropriate volume-integral determines the mean value of the position. Other quantum theories should follow suit. This item pertains to the classical measurement of position. Clearly, classical measurement of position does not yield an absolutely precise value of the three spatial coordinates. Hence, the quantum definition of density is acceptable. In relativistic notation, particle density is the 0-component of a 4-vector, called the 4-current. Therefore, a theory of an elementary quantum particle should have a coherent expression for a 4-current.

CL.2 Interactions reveal further aspects of the particle's point-like attribute. For example, the classical Lagrangian of a charged particle is

$$L = -m\sqrt{1 - v^2} - ev^\mu A_\mu \qquad (3.4)$$

(see [3], p. 48). Here, Maxwell equations of the electromagnetic fields require a conserved 4-current of the charged particle (see [3], pp. 73-76):

$$j^\mu = ev^\mu, \quad j^\mu_{,\mu} = 0. \qquad (3.5)$$

Equations (3.4)-(3.5) demonstrate two facets of a quantum particle: Its wave properties should yield an expression for density that is the 0-component of a conserved 4-current, and its point-like attributes should show up in a collision of two elementary particles. Experiments substantiate these requirements. Thus, electron interference shows its wave properties and the deep inelastic electron-proton scattering shows that the electron and the proton's quarks behave as point-like particles (see [23], p. 270).

CL.3 The solution of classical equations of motion of a point-like particle contains a well-defined value of its momentum.

Quantum theories should determine a quantity that reduces to the classical momentum. Here, one can take advantage of the particle's energy-momentum 4-vector $(E, \boldsymbol{p})$. Therefore, a covariant structure of the theory and a consistent quantum definition of energy (see the next item) ensure a consistent momentum definition.

CL.4 The solution of a classical equation of motion contains a well-defined value of its energy. Quantum theories should define an appropriate expression for the Hamiltonian, and eigenvalues of this Hamiltonian should comply with the classical notion of energy.

3.3 The Variational Principle

Basic properties of the variational principle are explained below.

3.3.1 Basic Properties

Physical theories of particles take the form of an assembly of coherent mathematical structures. Differential equations and the variational principle are important elements of these theories. These elements are connected by the Euler-Lagrange equations, which show how the differential equations of a specific physical theory of a quantum particle are derived from an appropriate Lagrangian density using the variational principle. This book examines some specific physical theories of interacting particles and emphasizes the role of the variational principle and its associated differential equations.

Physics is an experimental science. A physical effect is recorded if the state of a measuring device changes with time. Therefore, a physical theory must describe the time evolution of an appropriate system. This conclusion holds if just one relevant experiment measures a time-dependent event. Mathematics describes such data using time-dependent differential equations. Hence, differential equations are a vital element of the physical theory of any particle. Moreover, SR shows the connection between space and time coordinates. This means that the differential equations are partial differential equations with respect to the space-time coordinates $(t, \boldsymbol{x})$.

Another feature of measurements requires that for any given particle, the differential equations and their Lagrangian density contain a coherent expression for the interaction of the examined particle with an external physical entity. This issue is discussed in section 3.8 on p. 38, and in other parts of this book.

The variational principle plays an important role in classical mechanics (see e.g. [21, 22], in relativistic classical electrodynamics (see e.g. [3]), and in general relativity (see e.g. [3]). Quantum theories are the main objective of this book. Indeed, the present structure of quantum theories depends on the variational principle. For example, Weinberg's textbook states that "all field theories used in current theories of elementary particles have Lagrangians of this form" (see [20], p. 300). The Noether theorem, discussed in section 3.4, is an important theoretical element of quantum theories because it connects between symmetries of the Lagrangian density and self-evident conservation laws. Indeed, "this represents a general feature of the canonical formalism, often referred to as Noether's theorem: symmetries imply conservation laws" (see [20], p. 307). This issue has valuable practical features because it relieves the theoretical work from the tedious task of proving that the theory abides by self-evident conservation laws, such as energy-momentum conservation and angular momentum conservation. However, a word of warning should be added here: The benefits of the variational principle may not hold if it uses a mathematically incoherent structure. This book proves that some of the physical theories accepted at the time of writing (2021) fail to comply with this requirement.

The general form of the Lagrangian density is

$$\mathcal{L}(\psi, \psi_{,\mu}, \psi^\dagger, \psi^\dagger_{,\mu}) = \sum_i (\psi^\dagger \hat{O}_i \psi), \qquad (3.6)$$

where the quantum function takes the form $\psi(x)$, x denotes the four space-time coordinates, and $\hat{O}_i$ denotes operators. For example, the Dirac Lagrangian density comprises operators like the γ^μ matrices, the derivative operator, the particle mass, and the external field. The form of the Lagrangian density (3.6) does not explicitly depend on x.

The Euler-Lagrange equations are derived from the Lagrangian density through the variational principle. The variation with respect to ψ is

$$
\begin{aligned}
\delta S &= \delta \int \mathcal{L}\, d^4x \\
&= \int \left[\frac{\partial \mathcal{L}}{\partial \psi} \delta \psi + \frac{\partial \mathcal{L}}{\partial \psi_{,\mu}} \delta(\psi_{,\mu}) \right] d^4x. \\
&= \int \left[\frac{\partial \mathcal{L}}{\partial \psi} \delta \psi - \partial_\mu \left(\frac{\partial \mathcal{L}}{\partial \psi_{,\mu}} \right) \delta \psi + \partial_\mu \left(\frac{\partial \mathcal{L}}{\partial \psi_{,\mu}} \delta \psi \right) \right] d^4x. \quad (3.7)
\end{aligned}
$$

The integral of the last term of (3.7) yields the values of the integrand at spatial infinity. It is assumed that at spatial infinity, $\psi = \delta \psi = 0$. Therefore, the last term of (3.7) can be removed. Equating the variation δS to zero and remembering that $\delta \psi$ is an arbitrary variation, one finds that the first and second terms in the last line of (3.7) are the Euler-Lagrange equations.

In a more general case, the Lagrangian density depends on more than one quantum function:

$$
\mathcal{L}(\psi_r, \partial \psi_r / \partial x^\mu, \psi_r^\dagger, \partial \psi_r^\dagger / \partial x^\mu), \quad (3.8)
$$

where the index r runs on all relevant functions. Here, each function is varied separately; thus, the variation of the rth function yields the Euler-Lagrange equation

$$
\frac{\partial \mathcal{L}}{\partial \psi_r} - \frac{\partial}{\partial x^\mu} \frac{\partial \mathcal{L}}{\partial (\partial \psi_r / \partial x^\mu)} = 0. \quad (3.9)
$$

One can use the form (3.7) of the action and derive the following requirements:

REC.1 In the units where $\hbar = c = 1$, d^4x has the dimension $[L^4]$, and the action S is dimensionless. Hence, every term of the Lagrangian density must have the dimension $[L^{-4}]$. In particular, the quantum function ψ acquires dimension.

REC.2 The action S and d^4x are Lorentz scalars; hence, every term of the Lagrangian density must be a Lorentz scalar. The right-hand side of (3.6) shows how mathematically complex quantum functions produce a mathematically real Lagrangian density. This is an important theoretical element, and this book proves that it applies to many specific SM theories.

3.3.2 The Next Steps

The form of the Lagrangian density (3.9) is certainly not the final word. As stated above, this book discusses quantum processes that are determined by strong, electromagnetic, and weak interactions. The different names of these interactions indicate that their specific Lagrangian densities must take different forms. Therefore, support for a given physical theory should be based on the following steps:

1. Each interaction should have a specific form of its Lagrangian density, and this form yields the corresponding Euler-Lagrange equations, which are the equations of motion of the systems. Like the Lagrangian density, the associated Euler-Lagrange equations are written in terms of the quantum function $\psi(t, \boldsymbol{x})$.

2. The equations of motion should be solved for specific physical systems, and the specific form of the quantum function $\psi(t, \boldsymbol{x})$ should be found.

3. Properties of the quantum function $\psi(t, \boldsymbol{x})$, which solves the theory's equations of motion, should be compared with experimental data. A successful result is regarded as support for the given theory.

An application of these points to some theories is discussed in section 3.6, on p. 28.

3.4 The Noether Theorem

The Noether theorem connects the invariance of the Lagrangian density under a given transformation and a conservation law. This is an important theorem that is discusses in textbooks. Below some consequences of this theorem are shown and new results are derived.

3.4.1 Properties of the Noether Theorem

Two significant examples are shown here. The form of the Lagrangian density is

$$\mathcal{L}(\psi, \psi_{,\mu}, \psi^\dagger, \psi^\dagger_{,\mu}) = \sum_i (\psi^\dagger \hat{O}_i \psi), \qquad (3.6)$$

where each of the quantum functions takes the form $\psi(x)$, x denotes the four space-time coordinates, and $\hat{O}_i$ denotes operators. As stated above, the Dirac Lagrangian density contains γ^μ matrices, a derivative operator, mass, and external fields. This Lagrangian density does not *explicitly* depend on x.

Let us put the Euler-Lagrange equation (3.9) of section 3.3 in this form

$$\frac{\partial \mathcal{L}}{\partial \psi} = \partial_\mu \frac{\partial \mathcal{L}}{\partial \psi_{,\mu}}. \tag{3.10}$$

Here, for the simplicity of the notation, a single function ψ is used. The partial derivative of the Lagrangian density with respect to x^μ is

$$
\begin{aligned}
\delta^\nu_\mu \frac{\partial \mathcal{L}}{\partial x^\nu} &= \frac{\partial \mathcal{L}}{\partial x^\mu} \\
&= \frac{\partial \mathcal{L}}{\partial \psi} \psi_{,\mu} + \frac{\partial \mathcal{L}}{\partial \psi_{,\nu}} \psi_{,\nu,\mu} \\
&= \left(\partial_\nu \frac{\partial \mathcal{L}}{\partial \psi_{,\nu}} \right) \psi_{,\mu} + \frac{\partial \mathcal{L}}{\partial \psi_{,\nu}} \psi_{,\nu,\mu} \\
&= \partial_\nu \left(\psi_{,\mu} \frac{\partial \mathcal{L}}{\partial \psi_{,\nu}} \right). \tag{3.11}
\end{aligned}
$$

A substitution of (3.10) yields the third line of this equation. The first and the last line of (3.11) yield

$$\partial_\nu \left(\psi_{,\mu} \frac{\partial \mathcal{L}}{\partial \psi_{,\nu}} - \mathcal{L}\delta^\nu_\mu \right) = 0. \tag{3.12}$$

The quantity inside the brackets is the energy-momentum tensor, and the vanishing 4-derivative proves that it conserves energy and momentum.

The invariance of the Lagrangian density (3.6) under rotation and boost means that the theory conserves angular momentum and that it takes a relativistic covariant form. The reader may prove this claim.

The right-hand side of (3.6) proves that the Lagrangian density is invariant under a global phase transformation of the quantum function $\psi \to \exp(i\alpha)\psi$, where α is a mathematically real constant. This property yields an expression for a conserved 4-current. Let us examine how the infinitesimal variation $(1 + i\alpha)$ affects (3.6). The invariance of $\mathcal{L}$ under the transformation means that it is unchanged by it. Because of their mathematically complex structures, the functions $\psi^\dagger, \psi$ are treated as independent variables, and the variation of ψ is examined. These arguments yield the following relations

$$
\begin{aligned}
0 &= \delta\mathcal{L} \\
&= \frac{\partial\mathcal{L}}{\partial\psi}\delta\psi + \frac{\partial\mathcal{L}}{\partial\psi_{,\mu}}\delta(\psi_{,\mu}) \\
&= \frac{\partial\mathcal{L}}{\partial\psi}(ia\psi) + \frac{\partial\mathcal{L}}{\partial\psi_{,\mu}}(ia\psi_{,\mu}) \\
&= ia\left(\frac{\partial\mathcal{L}}{\partial\psi} - \partial_\mu(\frac{\partial\mathcal{L}}{\partial\psi_{,\mu}})\right)\psi + ia\partial_\mu(\frac{\partial\mathcal{L}}{\partial\psi_{,\mu}}\psi)
\end{aligned}
\tag{3.13}
$$

The last line of (3.13) is obtained from a term that is added and subtracted. The Euler-Lagrange equations prove that the quantity inside the large brackets of the last line vanishes, and α is an independent variable. Therefore, (3.13) yields an expression for a conserved 4-current

$$
j^\mu = a\frac{\partial\mathcal{L}}{\partial\psi_{,\mu}}\psi,
\tag{3.14}
$$

where a is an appropriate coefficient. j^0 is the particle's density. Hence a is fixed so that $\int j^0 d^3x = 1$.

Let us put in words the meaning of this expression.

P.1 Terms of the Lagrangian density that are independent of derivatives $\psi_{,\mu}$ of the quantum function do not contribute to the 4-current.

P.2 Lagrangian density terms that are proportional to a derivative of the quantum function, contribute to the 4-current. This contribution is independent of derivatives.

P.3 Terms of the Lagrangian density that depend on a product of derivatives of the quantum function, contribute to the 4-current. This contribution comprises derivatives $\psi_{,\mu}$.

Let us see how these properties impose serious constraints on the structure of an acceptable Lagrangian density in general, and on that of an elementary charged particle, in particular. Chapter 3.2 shows that the Schroedinger theory provides an expression for a conserved 4-current. The density of this 4-current is used for a definition of an inner product of the Hilbert space of the theory. Thus, let $\rho(\psi^\dagger, \psi)$ denote density. In this case, the inner product is

$$
< \psi^\dagger|\psi >= \int \rho(\psi^\dagger, \psi)d^3r.
\tag{3.15}
$$

This inner product is required for the construction of the Hilbert space in QM. The correspondence between QFT and QM (see chapter 3.2) means that a consistent QFT requires a well-defined conserved 4-current. Moreover, Maxwellian electrodynamics also requires a conserved 4 current $j^\mu_{,\mu} = 0$ of an electrically charged particle. This 4-current is used in the interaction term of charges and electromagnetic fields

$$\mathcal{L}_{int} = -ej^\mu A_\mu, \qquad (3.16)$$

where e is the electric charge and A_μ is the 4-potential. The conserved 4-current is derived from the Noether theorem (3.14). If a term of the Lagrangian density belongs to category P.3, then the 4-current factor, j^μ, of the electromagnetic interaction term (3.16) depends on a derivative of the quantum function $\psi_{,\nu}$. This interaction term *contributes a new term* to the 4-current j^μ_{new}, and this depends on the external 4-potential A_μ. The result is a new interaction term where the dependence on the 4-potential is not linear. This is inconsistent with Maxwellian electrodynamics [3], because this theory requires that the electrodynamics interaction term depends linearly on the 4-potential A_μ (see, e.g., [3], section 30).

Landau and Lifshitz prove that the 4-current of an elementary classical particle is conserved (see [3], pp. 73-78). Hence, the Noether theorem provides a prescription for a conserved 4-current that abides by the correspondence principle.

3.4.2 Discussion

Several questions arise from the procedure where the Noether theorem is applied for a derivation of the 4-current of a given QFT of a particle.

Q.1 If the Lagrangian density is independent of derivatives of the quantum function $\psi_{,\mu}$, then this method does not yield a conserved 4-current. However, this issue avoids casting doubt on the meaning of the Noether theorem because in this case, the Euler-Lagrange equation (3.9) is not a differential equation. As explained in section 3.3, this is unacceptable: It means that every coherent QFT depends on derivatives.

Q.2 The derivation of the conserved 4-current depends on the invariance of the Lagrangian density under a global change of phase $\exp(i\alpha)$. This transformation does not hold for a mathematically real quantum function; hence, the procedure does not provide a 4-current for this kind of function. Nevertheless, this issue does not cast doubt on the meaning of the Noether theorem – as described later in this book, the de Broglie principle provides a proof of the unacceptability of a mathematically real function of a massive quantum particle (see section 4.2).

Q.3 The Noether theorem shows a method for deriving a conserved 4-current. However, it says nothing on the case where it fails to yield a coherent 4-current and such a 4-current is constructed using another method. This problem is not discussed in this book. However, there are some long-standing problems with inconsistent 4-currents that are derived from the Noether theorem; these problems have not been resolved for several decades. A good explanation for this state of affairs is that the Noether theorem covers all possibilities, and indeed, there is no coherent 4-current for any specific QFT that does not agree with the Noether expression (3.14).

> *Conclusion: If the Noether theorem uses a coherent Lagrangian density of a massive quantum particle then it ensures that the theory is consistent with conservation laws. Furthermore, it yields an important prescription for the construction of the particle's 4-current.*

Before this section is closed, it is interesting to note that the derivations of this section rely on the validity of the Euler-Lagrange equations.

3.5 The de Broglie Principle

This section outlines an argument that shows the benefit of the variational principle. The de Broglie principle is regarded as an element of quantum theories, and the principle says that a free quantum particle has wave properties where the phase is

$$\Phi \propto e^{i(\boldsymbol{k} \cdot \boldsymbol{x} - \omega t)}. \tag{3.17}$$

Here, $\boldsymbol{k}$ is the particle's momentum $\boldsymbol{p}$, and ω is its energy (see [24], pp. 119, 120; [18], pp. 3, 18). Experiments confirmed this principle for the electron a few years after its publication. Now, experiments have substantiated wave properties of elementary particles and some composite particles (see [25]).

Consider the power series expansion of the exponential function

$$e^{i\eta} = 1 + i\eta + ... \tag{3.18}$$

The pure number 1 is a dimensionless Lorentz scalar. It follows that η must be a dimensionless Lorentz scalar. This property of the de Broglie phase imposes a constraint on any quantum theory: *It must provide a dimensionless Lorentz scalar quantity that can be used for the de Broglie phase!* The phase $\mathbf{k} \cdot \mathbf{x} - \omega t$ of (3.17) satisfies this requirement because it is proportional to a contraction of two 4-vectors, $(t, \mathbf{x})$ and $(E, \mathbf{p})$, and the dimension of the energy-momentum 4-vector is $[L^{-1}]$.

The variational principle provides a convenient path to this goal. Let $\psi(x)$ denote a quantum function of a given particle and x denotes the four space-time coordinates. Thus, if the Lagrangian density $\mathcal{L}(\psi(x), \psi(x)_{,\mu})$ of a quantum theory is a mathematically real Lorentz scalar (see e.g. [20], p. 300) where the dimension is $[L^{-4}]$ then the action

$$S(\psi) = \int d^4 x \mathcal{L}(\psi(x), \psi(x)_{,\mu}) \tag{3.19}$$

is a dimensionless mathematically real Lorentz scalar. It means that the action of this Lagrangian density can be used for the de Broglie phase of a quantum system.

In many practical cases, the calculation is carried out in the system's rest frame where the overall momentum vanishes. Relying on the consistency of SR, one infers that it is enough to work in this frame and to find a covariant expression for the energy operator – namely, the Hamiltonian. For this end, one uses the Lagrangian density and derives the energy-momentum tensor

$$T^{\mu\nu} = \frac{\partial \mathcal{L}}{\partial(\psi_{,\nu})} g^{\alpha\mu} \psi_{,\alpha} - g^{\mu\nu} \mathcal{L} \tag{3.20}$$

(see [14], p. 310). The component T^{00} of this tensor is the Hamiltonian density

$$\mathcal{H} = T^{00} = \frac{\partial \mathcal{L}}{\partial \dot{\psi}} \dot{\psi} - \mathcal{L} \tag{3.21}$$

(see [26], p. 16).

These arguments prove that an application of a covariant expression for the Hamiltonian in the rest frame of the system ensures that the system abides by the de Broglie principle. It means that a de Broglie compatibility is analogous to that of conservation laws: An application of the variational principle to an appropriate

Lagrangian density relieves the theoretical work from the quite tedious task of proving that the quantum theory is consistent with
this requirement.

This argument is important for the bound state of several quantum particles (like the neutron that comprises quarks, etc.). As
mentioned above, this system abides by the de Broglie principle.
Hence, the overall action of the system can be used for a consistent
definition of the de Broglie phase. Another aspect of this discussion
indicates the vital role of the Hamiltonian in quantum theories.

Different arguments can be used for proving the role of the action in the definition of the phase of a quantum particle. This issue
can be found in the literature. For example, the correspondence
between QM and CPH leads to the same conclusion derived above
(see e.g. [24], section 32; [27], p. 20).

The de Broglie phase (3.17) indicates other quantum properties.
Here, the particle's energy and momentum are derived from the
operators

$$i\frac{\partial\psi}{\partial t} = E\psi, \quad -i\frac{\partial\psi}{\partial x} = p_x\psi. \tag{3.22}$$

As shown above, the Hamiltonian is the energy operator. Therefore, the energy expression of (3.22) yields the primary equation of
a massive quantum state

$$i\frac{\partial\psi}{\partial t} = H\psi. \tag{3.23}$$

This fundamental equation demonstrates the need for a coherent
expression for the *Hamiltonian operator.*

3.6 The Lagrangian Density and its Equations of Motion

This section outlines the significance of the variational principle
and its differential equations as important elements of physical theories of interacting particles. Several kinds of particles and their
interactions are described separately. This section is introduced
before specific cases are discussed in detail because this may help
readers acquire a better understanding of the general structure of
physical theories. Hence, it is recommended that readers may skip
some unclear details that are later explained in relevant chapters.
It is also recommended that readers reread this section after fin-
ishing the study of this book.

3.6.1 Classical Electrodynamics

The historical order of the theoretical progress of electrodynamics begins with specific mathematical formulas that describe relevant aspects of electromagnetic processes. In the 1860s, these formulas enabled Maxwell to construct a set of partial differential equations that describe the laws of electromagnetic fields called Maxwell equations (see. e.g. [28], pp. 217-219). The relativistic covariant form of these equations is (see [28], p. 551; [3], pp. 71, 79)

$$F^{\mu\nu}_{,\nu} = -4\pi j^{\mu}, \quad F^{*\mu\nu}_{,\nu} = 0, \tag{3.24}$$

where $F^{*\mu\nu} = \frac{1}{2}\epsilon^{\mu\nu\alpha\beta}F_{\alpha\beta}$ and j^{μ} is the 4-current of the electric charge.

Solutions of Maxwell equations (3.24) determine the time evolution of electromagnetic fields. These solutions depend on the 4-current of a massive charged particle j^{μ}. The Lorentz law of force shows how electromagnetic fields affect the time evolution of a charged particle. In terms of densities, this force is (see [3], p. 89)

$$\rho_m \frac{dv^{\mu}}{dt} = eF^{\mu\nu}j_{\nu}, \tag{3.25}$$

where ρ_m denotes mass density.

The Maxwell equations and the Lorentz law of force are the fundamental equations of classical electrodynamics. For example, Jackson stated that "when combined with the Lorentz force equation and Newton's second law of motion, these equations provide a complete description of the classical dynamics of interacting charged particles and electromagnetic fields" (see [28], p. 218).

Maxwell equations have extensive experimental support. For example, the results of Maxwell theory say that the photon is massless and chargeless. The experimental upper bound of the photon's mass is smaller by a factor of about 10^{-23} times the electronic mass, and the experimental upper bound of the absolute value of the photon's charge is smaller by a factor of about 10^{-35} times the absolute value of the electronic charge [29]. Furthermore, let ϵ denote the experimental deviation from the Coulomb law (which is embedded in Maxwell equations). Here, the electric field of a motionless charge takes the form

$$E = Q/r^{(2+\epsilon)}, \tag{3.26}$$

and the experimental upper bound of $|\epsilon|$ is negligible: $|\epsilon| < 10^{-16}$ (see table 2 in [30]).

SR was formulated about 40 years after Maxwell equations. This theory shows that Newtonian mechanics holds only for cases where the particles' velocity is much smaller than the speed of light. Thus, relativistic mechanics is a theory where the limit $v \to 0$ takes the form of Newtonian mechanics (see e.g. [3], p. 27).

Unlike the case of Newtonian mechanics, Maxwell equations inherently fit SR. Thus, they can be written in the relativistic covariant notation (3.24). The same is true with the Lorentz force (3.25).

The last theoretical step is the derivation of Maxwellian electrodynamics from the variational principle. The well-known textbook of Landau and Lifshitz [3] discusses this issue in detail.

The chronological order of the progress of classical electrodynamics is an example showing that physics is an experimental science, where the theory has developed based on experimental work that yielded formulas. Figure 3.3 describes the chronological order of the construction of the theoretical structure of classical electrodynamics mentioned above.

Let us define two electrodynamics theories. The theory that relies on the differential equations of Maxwell and the Lorentz force is called here Maxwell-Lorentz electrodynamics (MLE). MLE is discussed in [31], in the first 11 chapters of [28], and in many other textbooks. Theories of electrodynamics that are derived from the variational principle are called here variational electrodynamics (VE). A classical description of VE is discussed in [3]. Quantum aspects of VE can be found in every textbook on quantum electrodynamics (QED) and in every (or nearly every) QFT textbook.

It should be noted that MLE and VE are not identical theories. For example, MLE is independent of the electromagnetic 4-potential, whereas VE depends on this quantity.

3.6.2 Monopoles

Maxwellian electrodynamics says that monopoles do not exist in the universe. In mathematical terminology, Maxwell equation $\nabla \cdot \boldsymbol{B} = 0$ means that no monopole exists. In contrast, electric

Figure 3.3: *Chronological order of the construction of theoretical classical electrodynamics.*

fields and magnetic fields play a similar role in Maxwellian theory. For example, the tensor $F^{\mu\nu}$ of Maxwell equations (3.24) comprises electric and magnetic fields. Furthermore, in the case of radiation fields that are emitted from a unique source, the invariants of the electromagnetic fields take symmetric form

$$B^2 - E^2 = 0, \quad \boldsymbol{E} \cdot \boldsymbol{B} = 0 \tag{3.27}$$

(see [3], chapters 25, 47). The validity of Maxwellian electrodynamics means that it correctly describes experimental data. Hence, the previous arguments mean that they can be regarded as the experimental basis of this problem:

> *Problem A. What is the coherent extension of Maxwellian electrodynamics that describes the physical behavior of a system of charges and monopoles?*

The experimental fact of the absence of monopoles in the universe means that the route that is described in fig. 3.3 cannot be used for this assignment. Therefore, a resolution of Problem A should rely on theoretical arguments.

The first issue is the definition of a monopole. This task can readily be taken from the literature, where the duality transformation of electromagnetic fields is

$$\boldsymbol{E} \to \boldsymbol{B}, \quad \boldsymbol{B} \to -\boldsymbol{E}. \tag{3.28}$$

The corresponding charge-monopole transformation is

$$e \to g, \quad g \to -e, \tag{3.29}$$

where g denotes the unit of monopole strength. The standard literature regards these transformations as a monopole definition (see [28], pp. 251-252, [32], p. 1363). The duality transformations of the electromagnetic fields (3.28) can be put in tensorial form as

$$F^{\mu\nu} \to F^{*\mu\nu}, \quad F^{*\mu\nu} \to -F^{\mu\nu}. \tag{3.30}$$

The transformations (3.28) and (3.29) are sometimes called duality rotations by $\pi/2$ (see [28], p. 252).

It turns out that an application of the duality transformations (3.28)-(3.29) to the present form of Maxwellian electrodynamics yields a theory of monopoles and electromagnetic fields. This dual theory comprises no electric charge. The absence of electric charge means that Problem A is still open. Herein, the system obtained from these transformations is called the dual system.

Considering this problem, one may examine the following postulates:

P.1 Electromagnetic fields of the dual system are identical to electromagnetic fields of the ordinary Maxwellian electrodynamics.

P.2 The required resolution of Problem A is based on an appropriate extension of the Lagrangian density of the ordinary Maxwellian electrodynamics – namely, the Lagrangian density that is described in textbooks like [3].

A charge-monopole theory that is based on postulate P.1 was constructed by Dirac [33]. This theory contains strings of discontinuity that are connected to every monopole, and the trajectory of every charge is forbidden from crossing these strings. It can be concluded that this charge-monopole theory is inconsistent with the dual theory mentioned above because the dual theory has monopoles but no string of discontinuity. Another problematic point of this charge-monopole theory is that it says that the monopole strength g is quantized and its smallest value is $g^2 = 137/4$. This gigantic value casts quantum equations, such as the Dirac equation and the Schroedinger equation, into a mathematically uncontrollable state. Moreover, a large number of experimental attempts to detect a Dirac-like monopole have been carried out for many decades, and all these attempts have been in vain. For example, a recent report of the Particle Data Group states, "To date there have been no confirmed observations of exotic particles possessing magnetic charge" [34]. The theoretical problems of postulate P.1 and its systematic experimental failure cast serious doubt on the physical acceptability of this postulate. It is interesting to point out that more than 30 years ago, the author of this book predicted the failure of the monopole quest based on postulate P.1 [35]. This correct prediction was grounded on postulate P.2, whose consequences are described below.

Postulate P.2 was used for a construction of a Regular Charge-Monopole Theory (RCMT) [36, 37]. Details of this theory are presented in section 10.3 of chapter 10 which discusses strong interactions. The main results of this theory are as follows:

MR.1 Charges do not interact with bound fields of monopoles.

MR.2 Monopoles do not interact with bound fields of charges.

MR.3 Radiation fields of systems are identical, and both charges and monopoles interact with them.

MR.4 Unlike the case of the Dirac monopole theory [33], the strength of the RCMT elementary monopole unit g is a free parameter.

MR.5 Unlike the case of the Dirac monopole theory [33], the RCMT is free of string irregularities.

The real world provides a successful manifestation of properties MR.1-MR.4 of the RCMT. Assume that a quark carries a unit g of magnetic monopole and $g \gg e$. Thus, the RCMT monopoles dominate strong interaction processes. Experiments prove that an electron (a pure charge) does not participate in strong interactions (see, e.g., [38], p. 2). In contrast, an energetic real photon (radiation fields) interacts strongly with quarks. For example, "the limiting photon total cross-sections on neutrons and protons are nearly the same, indicating that the photon interaction does not depend primarily on the charge of the target" (see [39], p. 269). More details of the proton-neutron similarity of the hard photon scattering data are shown in Fig. 33, p. 296 of [39]. This effect is sometimes called *the hadronic properties of the photon*.

The experimental evidence excellently fits the outcome MR.1-MR.4 of the RCMT. Indeed, the success of the ordinary Maxwellian electrodynamics means that an electron carries an electric charge but no monopole. In accordance with this issue and the assumption that the monopole strength unit takes a value $g \gg e$, a successful description of the above-mentioned electron and hard photon data can be given. Although this effect has been known for more than half a century, SM textbooks that have been published in recent decades turned a blind eye to it. This state of affairs means that mainstream theories still provide no acceptable explanation for the strong interaction features of hard photons as of this writing in 2021.

The overwhelming advantage of the RCMT as an explanation of strong interaction effects is discussed in chapter 10. This issue is also discussed in [40–43]. Cases that are described in chapter 10 support the validity of the RCMT as a strong interaction theory. This section is dedicated to the merits of the variational principle and its equations of motion as theoretical elements that hold for

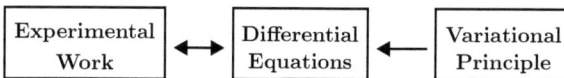

Figure 3.4: *Chronological order of the construction of the RCMT.*

many kinds of interactions. More details of specific interactions are discussed in separate chapters. Unlike the chronological order of the development of classical electrodynamics, shown in Fig. 3.3, the development of the RCMT takes a reverse chronological order, where the variational principle – which is a fundamental element of theoretical physics – provides a good starting point. This issue is illustrated in Fig. 3.4.

3.6.3 The Dirac Equation

The de Broglie principle was the first theoretical step to show the quantum features of a massive particle; this principle defines the relation between the momentum of a free quantum particle and its wavelength (3.17). A few years later, Schroedinger published his wave equation, which is the QM version of Newtonian mechanics (see [18], p. 19). An important justification for this equation relies on experimental data of atomic states. Generally, this equation describes some quantum properties of the electron. The Pauli equation, which accounts for the electronic spin effects, is an extension of the Schroedinger equation. The Dirac equation

$$i\frac{\partial \psi}{\partial t} = [\boldsymbol{\alpha} \cdot (-i\boldsymbol{\nabla} - e\boldsymbol{A}) + \beta m + eV]\psi \qquad (3.31)$$

was published soon after the Pauli equation [44]. In (3.31), $\boldsymbol{\alpha}, \beta$ are the ordinary Dirac matrices, whereas $(V, \boldsymbol{A})$ are the components of the electromagnetic 4-potential. This equation relies on relativistic arguments and *proves* the electron's spin degree of freedom, the value of its g-factor, the existence of the positron, and many other properties of spin-1/2 massive particles. Indeed, the Dirac equation "was validated by accounting for the fine details of the hydrogen spectrum in a completely rigorous way" [45]. An important quantity of the Dirac theory is the derivative-free conserved 4-current

$$j^\mu = \bar{\psi}\gamma^\mu\psi, \qquad (3.32)$$

where $\bar{\psi} = \psi^\dagger \gamma^0$.

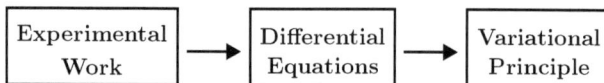

Experimental Work	→	Differential Equations	→	Variational Principle

Figure 3.5: Chronological order of the construction of the Dirac equation.

A later development shows that the Dirac equation (3.31) can be derived from a Lagrangian density that is now regarded as the Lagrangian density of QED (see e.g. [14], p. 78; [26], p. 84)

$$\mathcal{L}_{QED} = \bar{\psi}[\gamma^\mu(i\partial_\mu - eA_\mu) - m]\psi - \frac{1}{16\pi}F_{\mu\nu}F^{\mu\nu}. \qquad (3.33)$$

The term that contains A_μ is the interaction term of a charged Dirac particle with electromagnetic fields. Figure 3.5 describes the historical order of the progress of the Dirac theory of a massive spin-1/2 quantum particle. This is analogous to that of the Maxwell equations, as described in Fig. 3.3.

3.6.4 Weak Interactions

Chapter 11 includes a detailed discussion of experimental and theoretical aspects of weak interactions. In particular, that chapter refers to problems of the Lagrangian density of weak interactions and their equations of motion. At this point, the discussion is restricted to a primary problematic element of the SM sector of weak interactions, called the *electroweak theory*.

It turns out that *SM textbooks ignore the differential equations of the electroweak theory* [14, 46]). A fortiori, there is no indication that solutions of the missing electroweak equations fit experimental data. Moreover, many textbooks do not show the *full* form of the electroweak Lagrangian density. It is interesting to note that two textbooks and the present form (August 2021) of the Wikipedia electroweak item make some progress and write down an explicit form of the electroweak Lagrangian density [47–49]. This Lagrangian density is a cumbersome expression because it *comprises several dozens of terms*. (In contrast, the QED Lagrangian density (3.33) has just four terms!) These electroweak publications refrain from making the required step of showing the explicit form of the Euler-Lagrange equations and their solution. Furthermore, to the best of my knowledge, no SM textbook explains the reason for this grave discrepancy. Figure 3.6 depicts this unfortunate status of the electroweak theory. Chapter 11 discusses weak interaction problems in detail.

3.6.5 Summary

This section discusses the role of the variational principle and its differential equations as elements of the theoretical structure of particle theories. Electromagnetic, strong, and weak interactions

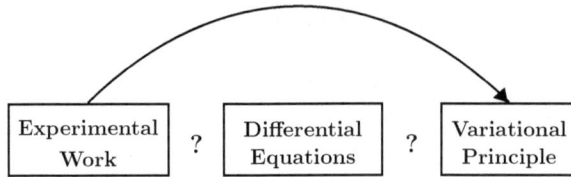

Figure 3.6: *Chronological order of the construction of the electroweak theory. The theory's differential equations are still absent.*

are briefly examined. Gravitation is the fourth kind of force that is known to physics. It is interesting to note that general relativity, which is the recognized theory of gravitation, can also be derived from the variational principle [3]. However, the gravitational force is outside the scope of this work. Few examples illustrate the primary role of the variational principle in theoretical physics. Later in this book, specific topics are discussed in detail.

As pointed out above, the variational principle relieves the theory of an elementary particle of three important tasks which are as follows:

1. *The de Broglie phase:* If the Lagrangian density is a mathematically real Lorentz scalar with a dimension of $[L^{-4}]$, then the action is a dimensionless Lorentz scalar that is suitable for the phase of the de Broglie wave.

2. *Conservation laws:* The Noether theorem indicates how a Lagrangian density provides a basis for writing down a theory that is consistent with conservation laws, like those of energy, momentum, and angular momentum.

3. *Charge conservation:* The Noether theorem shows how a Lagrangian density provides a prescription for the construction of a conserved 4-current, which is required by the GCP and Maxwellian electrodynamics.

It can be concluded that the variational principle and its Euler-Lagrange differential equations can be considered significant elements of coherent theories of elementary particles and their interactions. This approach is used in this book as the basis for an analysis of quantum systems.

3.7   Wigner's Analysis

Relying on mathematical arguments, it can be stated that Maxwell
equations are a harbinger of SR. Thus, Maxwell has recognized that
free electromagnetic waves are a solution of his equations. An important property of these waves is that they travel at the speed
of light in every inertial frame. This is a fundamental property of
SR. Einstein's formulation of SR was a breakthrough of the physical
concept about the structure of the universe. In this theory, physical
variables take the form of four-dimensional tensors that are defined
in Minkowski space. Lorentz transformations induce boosts and rotations in this space. These transformations make a six-parameter
group called the Lorentz group. An addition of space-time transla-
tions to the Lorentz group creates a 10-parameter group called the
Poincare group (also called the inhomogeneous Lorentz group).

Wigner has analyzed the unitary representations of the inhomogeneous Lorentz group [16]. His work demonstrates far-reaching
mathematical arguments that describe particle properties. Sternberg's description of the remarkable significance of Wigner's work
is as follows: "It is difficult to overestimate the importance of this
paper, which will certainly stand as one of the great intellectual
achievements of our century" (see [50], p. 149; see also [20], p. 1).

Besides Wigner's original work, his analysis can be found in
textbooks, such as in [50, 51]. Results of his work that are used
later in this book are as follows:

Wig.1 A quantum state has well-defined mass and spin.

Wig.2 There are two categories of physically meaningful systems:
The first category comprises systems with mass $m > 0$. The
velocity of members of such a system is smaller than the
speed of light. A member of this category has a well-defined
spin and its j_z takes $(2j + 1)$ different values. The second
category comprises massless particles, where $E^2 - p^2 = 0$ and
$E > 0$. They travel at the speed of light. Instead of spin,
these particles have 2 degrees of freedom of helicity.

Wigner's work is a purely mathematical analysis. It provides an
impressive example of the powerful merits of mathematics and the
physical relevance of results that are derived from its application.
His work uses fundamental elements of a quantum theory and SR,
which is a theory constructed on two principles – the principle of
relativity and the constant speed of light. One may wonder how can
these principles affect the quite irrelevant property of the number

of spin projections on the z-axis; because of this issue, it is quite surprising that Wigner has proved that J_z of massive particles has $(2J + 1)$ m-values. Nevertheless, a spin $= 1$ photon has only two values ± 1 of its polarization. Experiments support these claims.

Conclusions:

W.1 If two physical entities do not have the same mass, these entities are different. In particular, a massive particle and a massless particle are different physical entities.

W.2 If two physical entities do not have the same spin, these entities are different. In particular, if the spin projection of a spin ≥ 1 massive particle has more than two possible values. In contrast, the spin projection of a spin ≥ 1 massless particle has only two possible values. This means that a massive particle and a massless particle are inherently different physical entities even if they have the same spin.

W.3 The energy of a physical particle is greater than zero, and its energy-momentum invariant $E^2 - \boldsymbol{p}^2$ is non-negative. This means that the energy-momentum of a genuine physical particle cannot be space-like, and its energy should be greater than zero.

Unfortunately, Wigner's work is not adequately discussed in many QFT textbooks. Conclusions of Wigner's work are seriously examined in this book.

3.8   Measurability

Figures 3.3 - 3.6 show interrelations between experiments and theory in the historical progress of physics. Experiments are also used as acceptability criteria for a given theory: A theory must adequately describe experimental results that belong to its validity domain. These issues mean that the mathematical structure of a given theory must contain an element that shows how a physical effect arises. Here the Lagrangian density of electrodynamics can be used as a leading example (see (3.33) on p. 35). This expression comprises terms that depend on a single field – either the electron or the electromagnetic fields – and an interaction term that combines the two fields, ψ of the electron and the 4-potential A_μ of the electromagnetic fields:

$$\mathcal{L}_{Int} = -e\bar{\psi}\gamma^\mu A_\mu \psi. \tag{3.34}$$

As a part of the Lagrangian density, the interaction term must be a Lorentz scalar with a dimension of $[L^{-4}]$. The QED interaction term (3.34) abides by these requirements. This book uses this scheme for strong and weak interactions. It proves that the form of these interactions takes the simple structure of a single term that looks like the QED interaction term of (3.34) – a factor like e that represents the strength of the interaction and a Lorentz contraction of an expression of the Dirac γ^μ matrices with a tensorial expression of an external field.

The results of the Wigner analysis of section 3.7 are instructive elements on how to construct the required interaction terms. Wigner's work says that a massive quantum particle is characterized by its mass (which is a Lorentz scalar) and its spin. The QED interaction uses the electric charge, which – like the mass – is a Lorentz scalar. This term takes the form of a contraction of one 4-vector of γ^μ matrices with the electromagnetic 4-potential. This book shows that besides electromagnetic interactions, the form of this term is suitable for strong interactions. The spin of a Dirac particle is related to $\sigma_{\mu\nu}$ of (11.3), which is an antisymmetric product of two 4-vectors of the Dirac γ^μ matrices. Pauli was the first person who has examined this kind of interaction [52]. This book relies on the mathematical properties of the Pauli term and proves that it pertains to weak interactions.

> Conclusion: A theory of a quantum particle should contain a mathematical expression that describes how the time evolution of the particle is related to a physical effect. Expressions like the QED interaction term of the theory's Lagrangian density (3.34) can be used for this end.

3.9 A Summary of Constraints

Constraints on physical theories are extremely useful elements of the structure of theoretical physics. They provide necessary conditions, and every acceptable physical idea should comply with them. Obviously, every once in a while the compatibility of every constraint should be reexamined. For example, people thought that a physical object can have either particle *or* wave properties, but not both. This opinion was the basis for the Newton-Huygens differences about the nature of light. The distinction between these concepts has lasted for more than 200 years until Einstein's work on the photoelectric effect and the rise of QM. A proof of the logical compatibility between these notions is given in section 7.1.

Several constraints that are imposed on a quantum theory are already discussed above, and they are listed in this section. Putting them in one place may help readers see the entire picture. Besides the constraints that are mentioned below, other constraints can be deduced from a mathematical analysis of the theories.

CNST.1 The compatibility with SR means that a quantum theory must have a covariant structure.

CNST.2 An acceptable quantum theory should be derivable from a Lagrangian density using the variational principle (see section 3.3). This Lagrangian density should be a mathematically real Lorentz scalar with a dimension of $[L^{-4}]$. The de Broglie principle and the Noether theorem (see section 3.4) justify this requirement.

CNST.3 A quantum theory should have well-defined partial differential equations (see section 3.6). These equations, which determine the time-evolution of the system, are the Euler-Lagrange equations of the theory's Lagrangian density (see CNST.2). Solutions to these equations should adequately describe experimental data.

CNST.4 The measurability requirement says that the Lagrangian density of an elementary quantum particle theory should have an interaction term (see section 3.3).

CNST.5 The de Broglie formula determines the wavelength of a free massive particle. It is related to the particle's momentum

$$\lambda = 2\pi\hbar/p. \tag{3.35}$$

It is proved below that not every SM QFT theory is consistent with this constraint.

CNST.6 Because of the GCP – specifically, the dog-tail constraint (see chapter 3.2) – the appropriate limit of a QFT of an elementary quantum particle should be consistent with QM and with CPH.

CNST.7 A quantum theory of a charged particle should have a coherent expression for a conserved 4-current $j^{\mu}_{,\mu} = 0$. The Noether theorem provides a prescription for the construction of this 4-current (see subsection 3.4.2).

CNST.8 The interaction term of the Lagrangian density of a charged particle should be proportional to its electric charge e.

CNST.9 Maxwell equations are independent of the 4-potential A_μ. However, VE uses A_μ as the coordinate of the electromagnetic Lagrangian density (see [3], section 30). Therefore, the Euler-Lagrange equations prove that no term of the Lagrangian density should have A_μ where the power is greater than unity.

CNST.10 Wigner's analysis (see section 3.7) proves that a massive quantum particle has well-defined mass and spin. Instead of spin, a massless quantum particle has helicity; its s_z can take only the two values $\pm s$.

CNST.11 Relationship CNST.10 yields an important result. Let a, b, denote numerical constants and ψ_1, ψ_2 be functions of two quantum particles; moreover, let $\psi_1 \neq \psi_2$. Such an expression as $\Phi = a\psi_1 + b\psi_2$ is unacceptable if these functions determine free quantum states that do not have the same mass or the same spin.

CNST.12 The self-mass of a physical particle can be positive or zero. Its energy is positive.

Chapter 4

Consequences of Constraints on Quantum Theories

Let us derive some results from the constraints listed in section 3.9. This discussion is restricted to general issues, and the implications of these constraints for quantum theories on specific particles are discussed later in this book.

A quantum theory of a massive particle should abide by several principles, as discussed above. This argument is briefly explained as follows:

P.1 The correspondence between quantum theories and CPH is discussed in chapter 3.2. This chapter explains why this principle applies to QM, RQM, and QFT. The principle says that the limit of the expectation value of a variable of a quantum theory should agree with the value of the corresponding variable of CPH. A prerequisite condition for a successful satisfaction of this issue is that a quantum theory should have an appropriate expression for every variable of the classical theory. Several requirements are derived from this correspondence. Thus, a theory of any massive quantum particle should define an expression for these quantities – energy, momentum, angular momentum, and the particle's position. The appropriate limit of these expressions should agree with the classical quantity.

The problem concerning the classical limit of QM was already solved in the early days of QM. Hence, it is enough to demand that for every quantum particle, the appropriate limit of physical quantities of QFT and RQM should agree with QM.

P.2 A quantum theory of a massive particle should abide by the de Broglie principle.

P.3 A quantum theory of a massive particle should abide by the results of the Wigner analysis.

Details of these points are discussed below.

4.1   Conservation Laws

CPH proves the conservation of energy, momentum, and angular momentum of a closed system. It follows that the classical limit of a quantum theory should abide by these laws. The common approach to this issue is to use the variational principle and the Noether theorem. These topics are discussed in sections 3.3 and 3.4.

Maxwell equations prove charge conservation; hence, a quantum theory of a charged particle should abide by this law. Like the case of the previous conservation laws, an application of sections 3.3 and 3.4 provides a prescription for substantiation of this requirement.

An important warning applies to these issues. An application of the variational principle and the Noether theorem may yield erroneous results if the theory's Lagrangian density takes an incoherent mathematical structure. This book discusses several examples of SM theories that demonstrate the importance of this warning.

4.2   The de Broglie Wave

The de Broglie hypothesis concerning the wave nature of a free massive particle is a fundamental principle of quantum theories (see section 3.5). This hypothesis says that the following relation holds between the particle's wavelength and its linear momentum:

$$\lambda = 2\pi\hbar/p. \tag{4.1}$$

The undulating properties of the particle's wave function can be written as a linear combination of these expressions (see [18], p. 18):

$$\sin(\boldsymbol{k} \cdot \boldsymbol{x} - \omega t), \quad \cos(\boldsymbol{k} \cdot \boldsymbol{x} - \omega t), \quad \exp \pm i(\boldsymbol{k} \cdot \boldsymbol{x} - \omega t). \tag{4.2}$$

(Please note that these expressions are not linearly independent.)

An application of this principle is as follows: A mathematically real function can be written as a linear combination of the first and second functions of (4.2). Hence, a mathematically real wave function of an elementary free massive particle moving along the x-direction takes the form

$$\psi(t, x) = A \sin(kx - \omega t - \delta), \qquad (4.3)$$

where A and δ are real constants. The free elementary quantum particle analyzed herein is massive, and it has a rest frame. In this frame, the particle's linear momentum is $p = k = 0$, and its wave function (4.3) reduces to the form

$$\psi(t, x) = A \sin(-\omega t - \delta). \qquad (4.4)$$

Therefore, for every integer n, the mathematically real wave function (4.4) vanishes identically throughout the entire three-dimensional space at every instant t when $\omega t + \delta = n\pi$. This result means that the particle disappears at these instants. Therefore, no conserved expression for density can be defined for a mathematically real quantum function. This conclusion is consistent with the Noether expression for density (3.14). Indeed, (3.14) is based on the invariance of the Lagrangian density with respect to multiplication by the *complex factor* $\exp(i\alpha)$. However, this complex factor is unacceptable for a mathematically real function. An analogous argument can be found in the textbook of Berestetskii, Lifshitz and Pitaevskii [53] (pp. 42, 43). A different argument proving the need for a mathematically complex wave function is shown in Merzbacher' textbook: "...we are thus led to the conclusion that ψ waves describing free particle motion must be complex quantities..." [15] (pp. 14-16).

> Conclusion: A mathematically real function cannot describe a massive quantum particle.

The results show that the electroweak theory of the Z particle, the mathematically real version of the Higgs boson, the Majorana neutrino, the mathematically real version of the Klein-Gordon (KG) particle, and the Proca theory of a massive photon must be wrong. Other specific arguments that prove these results are shown later in this book.

4.3 The Correspondence between QM and CPH

The standard definition of the correspondence principle covers several aspects of the correspondence between QM and CPH, and QM textbooks discuss them (see e.g. [18], pp. 25-27, 135-138). However, some topics are apparently not well known.

4.3.1 The Point-like Attribute of an Elementary Particle

Landau and Lifshitz prove that a relativistic classical theory requires that an *elementary particle* must be point-like (see [3], pp. 46, 47). One of their arguments is as follows: By definition, an elementary particle does not comprise several distinct components that may move with respect to one another. Hence, if such a particle is not point-like, it must be absolutely rigid. Assume that this particle is hit on its left side and starts moving rightwards. Its absolute rigidity means that all spatial elements of this particle share the same acceleration at every instant. Hence, the applied force propagates at an infinite speed inside the volume of this particle. This is inconsistent with SR. For this reason, it is concluded that "within the framework of classical theory elementary particles must be treated as points" (see [3], p. 47).

Let us examine this topic in the quantum domain. This theory uses a function $\psi(x)$ for a description of an elementary particle, and the field's equations are the Euler-Lagrange equations of an appropriate Lagrangian density whose general form is

$$\mathcal{L}(\psi(x), \psi(x)_{,\mu}). \tag{4.5}$$

Here $x \equiv (t, \boldsymbol{x})$ denotes the four space-time coordinates. QFT textbooks support this approach. For example, Weinberg asserts that "All field theories used in current theories of elementary particles have Lagrangians of this form" (see [20], p. 300). He also states that the form of the Lagrangian density (4.5) refers to an *elementary particle*.

Let us examine the point-like properties of the quantum function $\psi(x)$ of the Lagrangian density (4.5). This function depends on a single set of four space-time coordinates, meaning that $\psi(x)$ can describe the probability of finding the particle at the space-time point x but *cannot* describe the distribution of the particle around x. Hence, the form of the QFT Lagrangian density (4.5) applies to an elementary point-like particle.

The foregoing analysis yields two important results:

1. Elementary particles of CPH and quantum physics share the same point-like form. This is an example that shows the consistency of quantum theories that use a Lagrangian density of the form (4.5), where the classical limit of QM agrees with CPH.

2. Experimental data support the distinction between elementary particles and composite particles. The electron and nucleon data show this point. It is recognized that nucleons are composite particles where the state is determined by three valence quarks. Nucleons have a finite size, and their radius is about 1 fm (see [8], p. 12). By contrast, an electron has the properties of an elementary point-like particle. Experiments have not measured a nonvanishing value of the electron's radius, and the experimental upper bound of this radius is about seven orders of magnitude smaller than the proton's radius [54].

   Another issue that demonstrates the difference between elementary particles and composite particles is the level of accuracy of calculations of the hydrogen atomic states and the chemical state of molecules. Here, the fundamental force and the theory are the same, representing QM states that are determined by the laws of electrodynamics. However, states of the hydrogen atom are calculated quite accurately [55]. These calculations treat the electron as a point-like quantum particle. By contrast, "molecules are considerably more complex in structure than atoms, and correspondingly less has been accomplished in the quantitative application of quantum mechanics to molecular problems" (see [18], p. 298).

This is an important conclusion because, at present (August 2021), the point-like attribute of an elementary particle is apparently not well known. An example that supports this claim is discussed in subsection 10.5.3 on p. 132.

4.3.2 The Principle of Complementarity

It is proved in section 4.3.1 that an elementary particle is point-like. Experiments support this outcome. On the other hand, section 3.5 explains the de Broglie principle, where a massive quantum particle has wave properties. A point-like particle is located at a spatial point whereas a wave is spread over a nonvanishing spatial region.

These properties *look* mutually contradictory. This issue is stated in textbooks. For example: "The existence of the wave-corpuscle duality is incompatible with classical doctrine" (see [56], p. 21). Similarly, the introduction of the principle of complementarity aims to bridge this gap: "Bohr, who was guided by the empirical fact of the dual nature of matter, has elevated this program to a *principle of complementarity*. According to this principle, wave and particle nature are considered complementary aspects of matter, both equally essential for a full description of the phenomena" (see [15], p. 7).

The idea that point-like particle and wave are inconsistent properties of a physical object can be found in many places [57]. Another aspect of this issue is that fields (that satisfy a wave equation) and particles are distinct concepts (see [20], p. 1). However, this subsection *proves* that point-like and wave are just two *mutually compatible* properties of the classical and quantum theoretical structures that are described above.

Examples of classical waves, like a wave in a pond or a sound wave, are a manifestation of the motion of elements of a medium. Here the system's energy is distributed throughout a quite large spatial region of the medium's components. Hence, people who adhere to concepts of CPH may regard the point-like and the wave properties of a quantum particle as a contradiction. However, this conclusion is unjustified. Indeed, an analogy with many other cases indicates that the scientific foundation of the wave phenomenon relies not on its human interpretation but on *the mathematical structure of the wave equation!* The wave equation illustrates this issue:

$$\nabla^2 \phi = \frac{1}{v_g^2} \frac{\partial^2 \phi}{\partial t^2}, \tag{4.6}$$

(see [3], p. 117; [58], p. 5). This equation depends on the group velocity v_g, and *it is independent of any medium*. In the previous examples of CPH, properties of the medium determine the velocity v, and in a quantum theory, v is determined by the de Broglie expression (3.17), which depends on the particle's energy and momentum. (Note that the relativistic relations $p^2 + m^2 = E^2$ of a massive particle yield $p\,dp = E\,dE$. Here, the product of the phase velocity and the group velocity is $v_g v_\phi = c^2 = 1$.) Furthermore, it is shown above that the quantum function $\psi(x)$ describes a point-like particle, *and at the same time*, an expression of a free quantum particle satisfies the wave equation. These arguments prove that the particle-wave properties of a quantum particle are consistent elements of quantum theories. *This result proves the principle of complementarity*. Namely, the principle of complementarity is not a postulate that is required for the coherent logical structure of quantum theories.

This outcome has strong experimental support. Take, for example, the double-slit interference experiment [59]. Here an appropriate beam of electrons (or another kind of a quantum particle) is split into two sub-beams that pass through two slits and produce an interference pattern on a screen. Each electron behaves as a wave that exists at the two spatially distinct regions of the sub-beams. Later, the interference pattern is generated as an assembly of distinct dots, where each of which results from a single electron that hits the screen. Here, the interference pattern describes the wave property, and each dot describes the particle's point-like property.

The examples of a classical wave, like a wave in a pond or a sound wave, depend on a medium by means of which the wave phenomenon is manifested. Here, the energy of the process is distributed throughout an appropriate portion of the medium – namely, a multiparticle object. This effect looks contradictory to the particle concept. However, electromagnetic waves are an important example of classical waves that are independent of a medium. Here, Maxwell equations define the wave velocity c that is used in the wave equation. In this case, the group velocity and the phase velocity are equal. It is explained above why the independence of the wave equation (4.6) on the existence of a medium enables reconciliation between the particle and the wave notions.

The confirmation of the de Broglie hypothesis of the wave nature of a quantum particle adds another blow to the ether concept. Indeed, the ether assumption means that in addition to the photon's ether concept, one requires the addition of a different kind of ether for every kind of quantum object, like the electron, the muon, and even the neutron. This is a strange combination of several kinds of ether, which has no support in the scientific literature.

Sections 4.2 and 4.3.1 explain how the quantum function $\psi(x)$ satisfies wave properties and point-like properties, respectively. Therefore, in quantum physics, there is no conflict between the notions of wave and a point-like particle. The same result holds for classical physics of electromagnetic radiation. Here Maxwell equations *prove* the existence of electromagnetic waves that travel at the speed of light. These attributes are independent of any medium and are compatible with Einstein's corpuscular explanation of the photoelectric effect.

4.3.3 The Inherent Uncertainty of Quantum Theories

A solution of the equations of motion of a classical particle determines its dynamical variables. If the canonical equations of the Hamiltonian are used, these solutions determine the particle's position and momentum. Hence, the classical limit of QM that is described in chapter 3.2 says that a quantum theory should define the particle's position and momentum, and the classical limit of these expressions should agree with the classical values, respectively.

Variables of CPH take a mathematically real value. Analogously, the value of a dynamical variable of a quantum particle is derived from a Hermitian operator where the eigenvalues are mathematically real quantities (see [20], pp. 49, 50).

A mathematical property of Hermitian operators is that not every two Hermitian operators commute. As an example, let us examine the commutation relations between the x coordinate and its conjugate momentum. These relations and the associated uncertainty are

$$xp_x - p_x x = i \quad \to \quad \Delta x \, \Delta p_x \geq 1. \tag{4.7}$$

(see [24], p. 87). Relations (4.7) are well documented in textbooks: "The Heisenberg uncertainty relation is thus seen to be a direct consequence of the noncommutativity of the position and the momentum operators" (see [15], p. 160).

The relativistic covariant form of the theory yields the corresponding energy-time uncertainty relations, where the Hamiltonian H is the energy operator

$$Ht - tH = i \quad \to \quad \Delta E \, \Delta t \geq 1. \tag{4.8}$$

The uncertainty relations between quantum operators mean that the determinism of CPH does not hold in quantum theories. The following argument proves the indeterministic aspect of quantum theories.

Consider a classical theory of a system of n elementary particles that relies on a given Lagrangian

$$L(q_i, \dot{q}_i, t) \tag{4.9}$$

(see e.g. [22], p. 35; [21], p. 2). This Lagrangian depends on the particles' coordinates, their time-derivatives, and the time. The Legendre transformation casts this Lagrangian into the Hamiltonian

$$H(q_i, p_i, t) \tag{4.10}$$

(see e.g. [22], p. 337; [21], p. 131). This Hamiltonian depends on the coordinates, their conjugate momenta, and the time. The theory based on the Lagrangian (4.9) is equivalent to the theory based on the Hamiltonian (4.10).

In the case of a closed system, the deterministic attribute of CPH says that *if the initial values of the coordinates and the momenta are accurately given at time t_0, the theory that is based on the Hamiltonian (4.10) provides accurate values of the system's variables at any given time t.* Evidently, this goal cannot be achieved in quantum theories, simply because the coordinate-momentum uncertainty relation (4.8) prevents an accurate determination of these variables at any given instant. It means that quantum theories should be based on different principles, and they cannot yield the deterministic solution of the differential equations that are derived from the classical Hamiltonian (4.10). The latter equations depend on the accurate value of the coordinates and momenta at a given time. Alternatively, quantum theories are bases on the action whose Lagrangian density (3.19) depends on the quantum function and its space-time derivatives $\psi(x)$, $\psi(x)_{,\mu}$. The function $\psi(x)$ depends on the space-time coordinates $(t, \boldsymbol{x})$, and the Lagrangian density of (3.19) *indirectly* depends on these coordinates. However, the correspondence principle that is described in section 3.2 proves the logical compatibility of quantum theories with classical physics.

Moreover, the deterministic attribute of CPH *cannot be realized.* Indeed, it depends on precise values of the position and the momentum of the Hamiltonian (4.10) at a given time t_0. Because of the uncertainty of QM, precise values of these variables cannot be determined in CPH.

4.4 The QFT and QM Correspondence

QFT takes a higher hierarchical rank with respect to QM, and the Schroedinger equation is the fundamental equation of QM. Relationships between these theories are examples of the dog-tail concept of chapter 3.2. This means that QM and the Schroedinger equation impose constraints on the structure of every specific QFT theory. This section is dedicated to this topic.

CPH imposes constraints on QM, and QM transfers these constraints to QFT. In particular, QFT should provide coherent expressions for energy, momentum, angular momentum, and density, where the limit should agree with QM. Special attention is devoted to the Noether theorem in general and the examination of its results in particular.

4.4.1 A Mathematically Complex Wave Function

It was shown in section 4.2 that the Schroedinger wave function must take a mathematically complex form. This attribute casts doubt on the compatibility of a mathematically real quantum function that describes a massive particle: Can the appropriate limit of a physical variable that is described by a mathematically real quantum function fit the value obtained by QM? A simple example that proves that the answer is negative is as follows: The Schroedinger theory provides an expression for a conserved 4-current whose density is $\psi^*\psi$ (see [15], p. 37). Therefore, every QFT theory of a massive particle must provide an expression whose limit is a conserved 4-current.

Section 4.2 gives a proof of the inability to define a consistent expression for density if the quantum function is mathematically real.

> Conclusion: A quantum theory of a massive particle that is written in terms of a mathematically real function violates the correspondence between QFT and QM.

4.4.2 The Hilbert and Fock Spaces

Section 4.4.1 shows that the Schroedinger equation of QM provides an expression for the quantum particle's density

$$\rho = \psi^*\psi. \tag{4.11}$$

In particular, $\psi \neq 0$ somewhere. It means that $\rho \geq 0$, and it does not vanish identically. This property enables a definition of an inner product for a Hilbert space

$$<\psi_1^*|\psi_2> \equiv \int \psi_1^*\psi_2 d^3r. \tag{4.12}$$

A convenient basis for the Hilbert space comprises orthonormalized functions, where

$$<\psi_i^*|\psi_j> = \delta_i^j. \tag{4.13}$$

A mathematical procedure called Gram-Schmidt [60] shows how to construct these functions.

The Hilbert space is an element of the mathematical structure of QM (see e.g. [20], pp. 49, 50; [56], pp. 164-166). This space is an important mathematical tool for the calculation of quantities that belong to the validity domain of QM. Its basis comprises single-particle wave functions that are useful for the single-particle Schroedinger and Dirac theories. The orthonormalized Hilbert space enables a simpler calculation of matrix elements that represents an operator $\hat{O}$

$$<\hat{O}>_{ij}=<\psi_i^*|\hat{O}|\psi_j>=\int \psi_i^*\hat{O}\psi_j d^3r, \qquad (4.14)$$

where ψ_i, ψ_j belong to the Hilbert space basis. In particular, the eigenfunctions of the Hamiltonian make a useful basis for the Hilbert space. These eigenfunctions represent physically meaningful states that are directly related to experimental measurements. In this case, appropriate operators enable the calculation of transition rates between energy states. This issue is an important connection between theory and experiment, which is a crucial property of a valid theory. In other words, Eq. (4.14) shows how the quantum functions ψ_i, ψ_j of the Hilbert space are used for explaining the connection of operators, which are mathematical objects, to the physical world.

Evidently, the Hilbert space is unsuitable for QFT, because this theory also describes states where the number of particles is not well defined. For example, experiments prove that the proton comprises 3 *uud* valence quarks and a probability of $\bar{q}q$ pairs of quarks of the u, d, s flavors [61]. Hence, QM and RQM are unsuitable for a calculation of the proton's state.

This problem is settled by enhancing the Hilbert space to the Fock space (see e.g. [47] pp. 40-42; [62]). The Fock space of n particles is produced from n Hilbert spaces.

> *Conclusion: The Hilbert and the Fock spaces are vital elements of the mathematical structure of quantum theories, because they enable the calculation of matrix elements of quantum operators.*

Unfortunately, it turns out that many SM textbooks refrain from a discussion of the Fock space. Furthermore, if these textbooks speak on operators, then the meaning of these operators is vague.

Chapter 5

The Noether Theorem and the Energy-Momentum Tensor

This chapter shows the significance of the energy-momentum tensor of physical fields. The Noether theorem shows how this tensor can be derived from the Lagrangian density of a given field. This chapter proves that the energy-momentum tensor can also be used for a consistency test of a field theory. This matter is (probably) new. The results show that the Dirac Lagrangian density of a spin-1/2 massive particle yields consistent results. On the other hand, problems exist with the present structure of electromagnetic fields in QED, and with quantum fields of massive particles that are described by a second-order differential equation. The problematic aspect of each of these issues is confirmed by independent analysis.

5.1 Introduction

The following example shows the role of the energy-momentum tensor as an element of a coherent structure of a physical theory. In the 1960s, Shockley and James presented a paradox where a stationary system of a charge and a magnet has a non-zero electromagnetic linear momentum [63]. They coined the term "hidden momentum" for a description of the missing momentum. Momentum is a fundamental element of physics and the somewhat mysterious "hidden momentum" concept indicates an unsettled problem. Soon after the publication of this paradox, Coleman and Van Vleck provided a general proof showing that the system's total linear momentum must be balanced [64]. Their analysis relies on the conservation properties of the energy-momentum tensor. Later, the author of this book analyzed the energy-momentum tensor of the Shockley and James classical system [65]. This analysis proves that an explicit mechanical linear momentum exists in the system. In particular, if a nonvanishing pressure gradient exists along a closed loop of current, effects related to the energy-momentum tensor yield a nonzero mechanical linear momentum. This mechanical momentum balances the electromagnetic linear momentum and supports the validity of the Coleman and Van Vleck general analysis. Thus, the Shockley and James paradox is an example that demonstrates the crucial role of the energy-momentum tensor as an element of theoretical physics.

The main objective of this chapter is to show that relations between the Lagrangian density of a physical field theory and its energy-momentum tensor can be used as a tool for an examination of the consistency of this theory. This is a new feature of the energy-momentum tensor, and the validity of the results is confirmed by independent analysis. Obviously, nobody openly denies that error removal takes a top priority of any scientific endeavor!

5.2   Relevant Principles

The following principles are used in the analysis carried out herein:

P.1 The variational principle and its associated Lagrangian density are regarded as fundamental elements of the present structure of field theory (see chapter 3).

P.2 Wigner's analysis of the unitary representations of the inhomogeneous Lorentz group proves that a quantum particle is characterized by mass and spin. A massless particle has two components of helicity [16, 50, 51]. This means that a quantum theory must provide a consistent expression for angular momentum. Furthermore, physical particles belong to one of the following categories: massive particles whose 4-

momentum is time-like and massless particles that have a null 4-momentum, where $E^2 - p^2 = 0$ and $E > 0$.

P.3 The correspondence between physical theories is discussed in chapter 3.2.

These principles provide constraints that apply to the acceptability of a physical theory. They are denoted by $P.n$, where n denotes one of the principles mentioned above. Other principles that are not mentioned above are also used, and they are mentioned below in appropriate places.

5.3 Properties of the Energy-Momentum Tensor

The variational principle P.1 guarantees the existence of many important properties of physical equations. This chapter examines the relevance of this principle to the construction of the energy-momentum tensor of some field theories.

Density has the dimension $[L^{-3}]$ and energy has the dimension $[L^{-1}]$. Therefore, energy density has the dimension $[L^{-4}]$. Furthermore, density is the 0-component of a 4-current (see [3], p. 75), and energy is the 0-component of the energy-momentum 4-vector (see [3], p. 29). Therefore, energy density is the component T^{00} of a second rank tensor $T^{\mu\nu}$, whose dimension is $[L^{-4}]$. This tensor is called the energy-momentum tensor. Entries of this tensor have a physical meaning (see [3], pp. 82-85). For example, the four entries $T^{\mu0}$ represent energy-momentum density, the four entries $T^{0\nu}$ represent energy 4-current, and for each i, the row $T^{i\nu}$ represents momentum 4-current.

The general practice of quantum theories analyzes the Hamiltonian equation $i\partial\psi/\partial t = H\psi$. This expression yields the energy of the quantum system. Item CL.3 on page 18 explains how momentum is consistently defined by the theory, provided it takes a relativistically coherent structure. Thus, the entries $T^{\mu0}$ of the energy-momentum tensor show the close energy-momentum relationships.

Consider the energy-momentum tensor of an elementary massive quantum particle in its rest frame. Isotropy of space means that entries that depend on a spatial direction must vanish. In particular, momentum-related entries and energy 3-current entries must vanish. (This conclusion also holds for an elementary quantum particle where the spin does not vanish. Indeed, spin is an axial vector, whereas 3-momentum and 3-current are polar vectors. Hence, spin cannot affect this conclusion.) It follows that the required tensor takes the following form:

$$T^{\mu\nu} = \begin{pmatrix} m\rho & 0 & 0 & 0 \\ 0 & 0 & 0 & 0 \\ 0 & 0 & 0 & 0 \\ 0 & 0 & 0 & 0 \end{pmatrix}, \tag{5.1}$$

where m denotes the particle's mass and ρ denotes its density. The tensor (5.1) is symmetric in the particle's rest frame. Hence, it is symmetric in all frames (see [66], p. 77).

As stated in item P.2, the Wigner analysis of the inhomogeneous Lorentz group shows that a quantum particle is characterized by mass and spin. Hence, the theory's structure must provide a well-defined expression for angular momentum. It turns out that a symmetric energy-momentum tensor is required for this end (see [3], pp. 82-85). This outcome also applies to fields that represent a massless particle, like the photon, which has no rest frame.

The Noether theorem shows that a Lagrangian density $\mathcal{L}(\psi, \psi_{,\mu})$, which does not depend explicitly on space-time coordinates, yields an expression for the energy-momentum tensor. This tensor takes the form

$$T^{\mu\nu} = \frac{\partial\mathcal{L}}{\partial\psi_{,\nu}}g^{\mu\alpha}\psi_{,\alpha} - g^{\mu\nu}\mathcal{L} \tag{5.2}$$

(see [3], p. 83; [14], p. 310). The tensor (5.2) satisfies energy-momentum conservation

$$T^{\mu\nu}_{,\nu} = 0. \tag{5.3}$$

The definition (5.2) is consistent with the required dimension of $T^{\mu\nu}$. Indeed, in units where $\hbar = 1$ the action S is dimensionless. Hence, the definition

$$S = \int \mathcal{L}\, d^4x \tag{5.4}$$

together with $c = 1$ prove that the dimension of the Lagrangian density and of its energy-momentum tensor (5.2) is $[L^{-4}]$.

The correspondence principle, P.3, says that the energy-momentum tensor of a classical body is relevant to an analysis of the energy-momentum tensor of quantum fields. In the rest frame of a classical macroscopic body, this tensor takes the form

$$T^{\mu\nu} = \begin{pmatrix} \epsilon & 0 & 0 & 0 \\ 0 & p & 0 & 0 \\ 0 & 0 & p & 0 \\ 0 & 0 & 0 & p \end{pmatrix}, \tag{5.5}$$

where ϵ denotes energy density and p denotes pressure (see [3], p. 92). However, for an elementary particle, the pressure vanishes and (5.5) reduces to (5.1). The discussion below examines the energy-momentum tensor (5.2) derived from the Lagrangian density of specific fields.

5.4 Maxwellian Fields

Textbooks on Maxwellian electrodynamics show the standard derivation of the electromagnetic energy-momentum tensor. Here, the Lagrangian density of free electromagnetic fields (see [3], p. 86; [28] p. 601)

$$\mathcal{L}_{EM} = -\frac{1}{16\pi} F^{\mu\nu} F_{\mu\nu} \tag{5.6}$$

is used. Evidently, this expression is a Lorentz scalar that does not explicitly depend on the space-time coordinates. Hence, the Noether theorem says that it should yield a conserved expression for energy-momentum. The calculation yields the following *non-symmetric* tensor (see [3], p. 86):

$$T^{\mu\nu} = -\frac{1}{4\pi}\frac{\partial A_\lambda}{\partial x_\mu} F^{\nu\lambda} + \frac{1}{16\pi} g^{\mu\nu} F_{\alpha\beta} F^{\alpha\beta} \tag{5.7}$$

This non-symmetric result means that something is wrong with the derivation of (5.7). Indeed, the calculation begins with the Lorentz scalar Lagrangian density (5.6), which does not depend explicitly on the space-time coordinates. In this case, the Noether theorem says that angular momentum should be conserved (see [26], pp. 18, 19). Obviously, if a theory conserves angular momentum, then it must provide a consistent expression for this quantity. However, section 5.3 shows that a consistent definition of angular momentum requires a *symmetric* energy-momentum tensor. Hence, the non-symmetric result (5.7) means that *something is wrong with its derivation*.

This outcome supports the discussion of chapter 8 about erroneous points in the present structure of electrodynamics. The specific reason for the problem discussed in that chapter is that radiation fields and bound fields are *different* physical entities. Hence, the tensor $F^{\mu\nu}$ of (5.6) should not be the sum of the tensors of these entities.

Relying on this conclusion, it seems important to find implications of a separate analysis of radiation fields and bound fields.

This analysis takes advantage of textbooks that show how the tensor (5.7) can be corrected (see [3], pp. 86, 87). Here, the following correction term

$$X^{\mu\nu} = \frac{1}{4\pi}\frac{\partial A^{\mu}}{\partial x^{\lambda}}F^{\nu\lambda} = \frac{1}{4\pi}\frac{\partial}{\partial x^{\lambda}}(A^{\mu}F^{\nu\lambda}). \qquad (5.8)$$

is added to (5.7), and the symmetric tensor

$$T^{\mu\nu} = \frac{1}{4\pi}\left(-F^{\mu\lambda}F^{\nu}_{\lambda} + \frac{1}{4}g^{\mu\nu}F^{\alpha\beta}F_{\alpha\beta}\right). \qquad (5.9)$$

is obtained.

Let us carry out a separate calculation of the correction tensor (5.8) of a monochromatic plane electromagnetic wave that moves in the z-direction and is linearly polarized in the x-direction. The 4-potential of this wave is discussed in the literature (see [3], pp. 118-120 or [28], pp. 269-278). In appropriate units, the physically meaningful mathematically real part of the 4-potential of this plane wave is

$$A_{\mu} = \cos(k(z-t))(0,1,0,0). \qquad (5.10)$$

(Note that in the units used herein, $c = 1$ and $(kz-\omega t) \to k(z-t)$.) The standard form of the electromagnetic field tensor is (see [3], p. 65)

$$F^{\mu\nu} = g^{\mu\alpha}g^{\nu\beta}(A_{\beta,\alpha} - A_{\alpha,\beta}) = \begin{pmatrix} 0 & -E_x & -E_y & -E_z \\ E_x & 0 & -B_z & B_y \\ E_y & B_z & 0 & -B_x \\ E_z & -B_y & B_x & 0 \end{pmatrix}. \qquad (5.11)$$

Let us calculate the correction term (5.8) in the case of the linearly polarized radiation fields discussed here. For these fields, one finds that in the units where $c = 1$, $|\boldsymbol{E}| = |\boldsymbol{B}|$ and only the components E_x and B_y do not vanish. Therefore, the tensor of the radiation fields discussed herein is ($|\boldsymbol{E}| = |\boldsymbol{B}| = 1$)

$$F^{\mu\lambda} = -k\sin(k(z-t))\begin{pmatrix} 0 & -1 & 0 & 0 \\ 1 & 0 & 0 & 1 \\ 0 & 0 & 0 & 0 \\ 0 & -1 & 0 & 0 \end{pmatrix}. \qquad (5.12)$$

The 4-potential (5.10) depends only on the coordinates z, t. Hence, a direct calculation of the partial derivatives of the 4-potential yields the tensor

$$\frac{\partial A^\mu}{\partial x^\lambda} = -k\sin(k(z-t)) \begin{pmatrix} 0 & 0 & 0 & 0 \\ -1 & 0 & 0 & 1 \\ 0 & 0 & 0 & 0 \\ 0 & 0 & 0 & 0 \end{pmatrix}. \tag{5.13}$$

A substitution of (5.12) and (5.13) into the correction term (5.8) proves that it is a null quantity. The generalization of this result to any radiation field is straightforward. It follows that in the case of radiation fields, the energy-momentum tensor (5.7), which is directly obtained from the variational principle, is already symmetric. Its form can be written in the well-known standard expression (see [3], p. 87)

$$T^{\mu\nu} = \frac{1}{4\pi}\left(-F^{\mu\lambda}F^\nu_\lambda + \frac{1}{4}g^{\mu\nu}F^{\alpha\beta}F_{\alpha\beta}\right). \tag{5.14}$$

Conclusions:

C.1 A symmetric energy-momentum tensor is directly obtained for radiation fields.

C.2 Radiation fields and bound fields are different physical entities.

C.3 There is a problem with bound fields as an independent physical entity (see chapter 8.2).

For further usage, let us express the energy density T^{00} (5.14) of electromagnetic fields as

$$\mathcal{W} = (E^2 + B^2)/8\pi. \tag{5.15}$$

The momentum density equals the energy current, and together, they are the $T^{0i} = T^{i0}$ entries of the energy-momentum tensor (5.14). This is the Poynting vector

$$\boldsymbol{S} = \boldsymbol{E} \times \boldsymbol{B}/4\pi \tag{5.16}$$

(see [3], p. 81).

5.5   The Dirac Fields

Let us apply the ordinary construction of the energy-momentum tensor (5.2) to a Dirac field. The Lagrangian density of a free Dirac field is (see [14], p. 52; [26], p. 54)

$$\mathcal{L} = \bar{\psi}(i\gamma^\mu \partial_\mu - m)\psi. \qquad (5.17)$$

Here, the dimension of the expression in the parentheses is $[L^{-1}]$. For this reason, the dimension of a Dirac function ψ is $[L^{-3/2}]$. With the general formula for the energy-momentum tensor (5.2), one finds

$$T^{\mu\nu} = \bar{\psi}i\gamma^\nu g^{\mu\alpha}\partial_\alpha\psi - g^{\mu\nu}\mathcal{L}. \qquad (5.18)$$

Let us use the definition $\bar{\psi} \equiv \psi^\dagger\gamma^0$ and examine entries of (5.18). Section 5.3 shows that the entry T^{00} is energy density. Here the expression (5.18) corresponds to the Legendre transformation, which casts the mechanical Lagrangian to the Hamiltonian (see [21], p. 131; [22], p. 337). This transformation removes terms that are proportional to the first-order time-derivative of the coordinates. The result is

$$T^{00} = \psi^\dagger(-i\boldsymbol{\alpha} \cdot \boldsymbol{\nabla} + \beta m)\psi, \qquad (5.19)$$

where $\gamma^0\boldsymbol{\gamma} = \boldsymbol{\alpha}$. The expression inside the parentheses of (5.19) is the Dirac Hamiltonian of a free particle (see [2], p. 11), and $\psi^\dagger\psi$ is the Dirac density. Hence, (5.19) represents energy density, in accordance with section 5.3.

Let us see the form of off-diagonal entries of (5.18) where $\mu \neq \nu$:

1. If $\mu = k > 0$ and $\nu = 0$ then $\gamma^0\gamma^0 = 1$ and (5.18) is

$$T^{k0} = \psi^\dagger(-i\partial_k)\psi. \qquad (5.20)$$

   Therefore, T^{k0} is the momentum density, in accordance with section 5.3.

2. If $\mu = 0$ and $\nu = k > 0$ then (5.18) is

$$T^{0k} = \psi^\dagger(i\alpha_k\partial_0)\psi. \qquad (5.21)$$

   Here, the Dirac α_k is the kth component of the velocity operator (see [2], p. 11). Hence, T^{0k} is the energy current. The relativistic 3-momentum is $\boldsymbol{p} = E\boldsymbol{v}$, where E is the energy and $\boldsymbol{v}$ is the 3-velocity. This means that $T^{\nu 0} = T^{0\nu}$. This relation is consistent with the required symmetry of this tensor, in accordance with section 5.3.

Other components of the energy-momentum tensor of the Dirac field can be calculated analogously.

> Conclusion: This section proves that an application of the standard construction of the energy-momentum tensor (5.2) to the Dirac field yields quantities that are consistent with physical requirements.

5.6 Second-Order Quantum Equations

An example of a Lagrangian density of a second-order quantum theory of a massive particle is

$$\mathcal{L} = g^{\mu\nu}\phi^*_{,\mu}\phi_{,\nu} - m^2\phi^*\phi + OT, \tag{5.22}$$

where OT denotes other terms. The first term on the right-hand side of (5.22) is a Lorentz scalar that is a contraction of two 4-gradients. A derivative-dependent Lorentz scalar can also take the form of a contraction of two 4-curls. The omission of the second case does not affect the discussion. In the case of the KG field, OT is null (see [20], p. 21; [26], p. 38). The sign of the mass term is irrelevant to the present discussion. Hence, the Lagrangian density (5.22) describes the appropriate expression for the Higgs boson (see [14], p. 715). An analogous expression holds for the electroweak $W^\pm$, Z bosons (see [47], p. 518) and for the Proca theory of a massive photon (see [28], pp. 597-601).

The $[L^{-4}]$ dimension of the Lagrangian density of a quantum field and a second-order quantum differential equation mean that the dimension of this field function ϕ is $[L^{-1}]$. For this reason, this kind of quantum theory is intrinsically different from the Dirac theory of a massive quantum particle, where the dimension of the field's function ψ is $[L^{-3/2}]$.

Let us utilize the standard construction of the energy-momentum tensor (5.2) and find an expression for energy density T^{00} of these fields. The mass-dependent term of (5.22) contains no derivatives. Therefore, in the corresponding energy-momentum tensor, this term takes the same form with the opposite sign

$$T^{00} = m^2\phi^*\phi + ... \tag{5.23}$$

This expression for energy density depends *quadratically* on mass. Therefore, it is inconsistent with the standard form of the energy-momentum tensor (5.1), as well as with the celebrated relativistic expression $E = mc^2$, both of which depend *linearly* on mass.

For this reason, the structure of the energy-momentum tensor of a second-order Lagrangian density of a massive particle demonstrates an inherent inconsistency of the associated theory. Other errors of second-order theories of massive particles are discussed in appropriate places in this book. These cases support the general principle stating that a physical theory that depends on an incoherent mathematical structure cannot successfully describe physical data.

5.7   Lessons Derived from the Tests by the Energy-Momentum Tensor

This chapter presents a novel application of the energy-momentum tensor: Its standard derivation from the Lagrangian density of a given theory can be used as an indication of the mathematical coherence of that theory. This approach relies on its mathematical structure, and the specific examples examined above justify its merits. This outcome is taken seriously in this book.

The Dirac theory of a spin-1/2 massive particle has successfully passed this test on the energy-momentum tensor successfully (see section 5.5 on p. 62). Hence, properties of the Dirac Lagrangian density deserve special attention. We can write it here in the following form:

$$\mathcal{L}_D = \bar{\psi}[\gamma^\mu i \partial_\mu - m - e\gamma^\mu A_\mu]\psi. \tag{5.24}$$

The quantities inside the square brackets are kinetic term, mass term, and an interaction term with the electromagnetic 4-potential. The interaction term of (5.24) is a contraction of the γ^μ 4-vector with the electromagnetic 4-potential.

Let us examine the Dirac Lagrangian density and point out its merits. The following analysis is restricted to the Dirac validity domain, where particles are not created or destroyed:

1. Experimentally, the Dirac equation is very successful. Solutions of its differential equation provide a very good description of the energy states of the hydrogen atom. It also explains the electronic g-factor. Peskin and Schroeder describe the merits of the Dirac electron: "That such a simple Lagrangian can account for nearly all observed phenomena from macroscopic scales down to 10^{-13} cm is rather astonishing" (see [14], p. 78).

2. A charged Dirac particle interacts linearly with the electromagnetic 4-potential. This is a mathematically simple form

and the interaction term of (5.24) is called *minimal interaction* (see e.g. [2], p. 11; [14], p. 78). Evidently, any specific interaction requires at least one term that is added to the Lagrangian density of a free particle. The Occam's razor principle supports theories that have this property.

Remark: it is stated in this book that gravitation is excluded from the discussion. General relativity does not use an interaction term, and it regards the metric as the dynamical variable. It means that the metric which is already used for gravitation cannot be used for other interactions that are discussed in this book. Hence, each of these interactions requires at least one specific term that is added to the Lagrangian density.

3. The Dirac γ^μ matrices are an important asset of the mathematical structure of the Dirac theory. Each of these matrices comprises dimensionless pure numbers, *and* the matrices transform like entries of a 4-vector. This 4-vector may be contracted with an external field and produce a *derivative-free* interaction term for the Lagrangian density. This is an important attribute of the Dirac theory because the Noether theorem proves that a derivative-dependent interaction term destroys the original definition of the 4-current (see 3.14 on p. 24). A theory of a massive quantum particle that has an integral spin lacks this benefit. Hence, to account for the interaction with a 4-vector or a 4-tensor field, a quantum theory of an integral spin particle introduces derivatives, like a 4-gradient or a 4-curl. It will be proved later in this book that inherent errors appear in interaction terms of integral spin theories of a massive particle, like the charged KG theory or the electroweak theory of the $W^\pm$ particles.

This book regards the success of the electromagnetic minimal interaction of the Dirac Lagrangian density (5.24) as a guiding light. It is shown below that strong and weak interaction theories can follow this course. According to the Occam's razor principle, these analogs of the electromagnetic minimal interaction are the preferred choice.

Chapter 6

The Noether Theorem and the 4-Current

The previous chapter showed how the Noether theorem provides a prescription for the construction of the energy-momentum-tensor and how this tensor can be used as a tool for testing the coherence of a field theory. This chapter examines an analogous aspect of the Noether theorem and shows that its expression for the 4-current of a quantum theory can be used for the same purpose. Several examples support this assertion.

The invariance of the Lagrangian density under a mathematically complex global phase transformation $\exp(i\alpha)$ (see (3.13) on p. 24) enables the Noether theorem to provide an expression for a conserved 4-current (see (3.14) on p. 24). In principle, this mathematically complex transformation is unsuitable for mathematically real quantum fields. Therefore, this chapter discusses theories that use mathematically complex quantum functions.

A crucial element of Maxwellian electrodynamics is the continuity equation, which describes a conserved 4-current of a charged particle. (see e.g. [28], pp. 217, 218). The analysis in this section examines mathematically complex functions that describe the time evolution of a system of a massive charged quantum particle and electromagnetic fields. The Noether theorem shows how a conserved 4-current can be constructed for this quantum theory. This conserved 4-current is required for Maxwellian electrodynamics, where the equations describe the time evolution of electromagnetic fields. Figure 6.1 illustrates the interrelations between these theoretical elements.

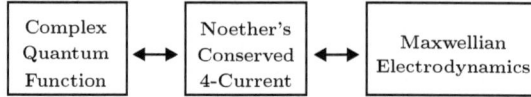

Figure 6.1: *Interrelations between three theoretical elements*

6.1 The 4-Current of the Dirac Field

Consider the QED Lagrangian density (see (3.33) on p. 35):

$$\mathcal{L}_{QED} = \bar{\psi}[\gamma^\mu(i\partial_\mu - eA_\mu) - m]\psi - \frac{1}{16\pi}F_{\mu\nu}F^{\mu\nu}, \qquad (3.33)$$

The QED Noether 4-current of the Dirac theory (3.33) (see (3.32) on p. 34) is well documented in textbooks as

$$j^\mu = \bar{\psi}\gamma^\mu\psi. \qquad (6.1)$$

This is a consistent derivative-free expression that has strong experimental support.

> *Conclusions: The Noether theorem yields a coherent expression for the 4-current of a Dirac particle. Experiments provide strong support for this expression. Furthermore, this outcome agrees with the coherent results of the analogous examination of the Noether expression of the Dirac energy-momentum tensor.*

6.2 The 4-Current of Fields of Second Order Quantum Equations

QFT textbooks discuss several kinds of Lagrangian densities of massive particles that yield second-order quantum equations: Here, the mathematically complex KG theory and the electroweak $W^\pm$ theories are examined.

The general form of the Lagrangian density of these theories is

$$\mathcal{L} = \eta m^2 \phi^\dagger \phi + OT, \qquad (6.2)$$

where η denotes a numerical constant, ϕ denotes the field function, and OT denotes other terms that account for the specific theory. The Lagrangian density has the dimension $[L^{-4}]$, and the mass has the dimension $[L^{-1}]$. Hence, the dimension of the quantum function ϕ is $[L^{-1}]$. The action and the Lagrangian density are mathematically real quantities. Hence, the mathematically complex function ϕ means that terms of the Lagrangian density that depend on ϕ should take the form of the product $\phi^\dagger \phi$. In addition, the 4-current is a mathematically real expression, and its dimension is $[L^{-3}]$. For these reasons, the 4-current of these functions must contain a derivative! However, a coherent 4-current of a charged particle should be derivative-free (see section 3.4). Hence, *there is no coherent electromagnetic interaction of a second-order quantum theory of mathematically complex functions of a charged particle.* (The Noether theorem yields the same result: The Lagrangian density (6.2) contains terms that have a product of two derivatives of the quantum function ϕ. Hence, the Noether 4-current has a derivative $\phi_{,\mu}$.) Specific discussions of the charged KG particles (see chapter 12.3 on p. 205) and the $W^\pm$ particles (see subsection 11.6.2 on p. 184) proves that after many decades that have elapsed since the formulation of these theories, they still have no coherent expression for their electromagnetic interaction.

> *Conclusion: The Noether theorem of a mathematically complex quantum function of an electrically charged particle does not necessarily provide a theoretically coherent expression for a conserved 4-current. This outcome applies to the charged KG particles and the electroweak description of the $W^\pm$ particles. Therefore, theories of these fields are inherently wrong because they cannot show a coherent electromagnetic interaction term.*

6.3   Density and Quantum Transitions

Subsection 4.4.2 on p. 52 explains why density is required for a coherent QFT structure. There is another reason for this requirement that pertains to a theoretical description of the decay of an elementary particle. As such, it has a close relationship with the need to explain experimental data.

Figure 6.2 illustrates the muon decay. This is a weak-interaction process where one Dirac particle is destroyed and three Dirac particles emerge. Let us examine the measurement of the outgoing particles of this figure. The process described in Fig. 6.2 cannot be measured in an ordinary laboratory because it is extremely hard to detect a single neutrino. Simultaneous detection of two neutrinos is *practically* impossible. However, this difficulty does not deny

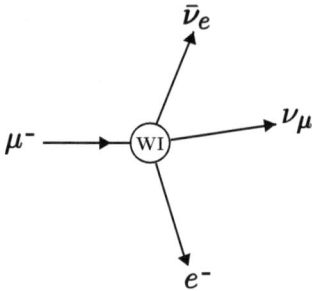

Figure 6.2: *The weak interaction decay of the μ^-.*

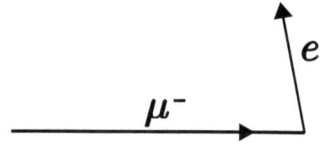

Figure 6.3: *The μ^- decay: The path of the μ^- and that of the outgoing electron.*

the theoretical analysis. Furthermore, the objective of the examination of the process of Fig. 6.2 is to prove that the experiment shows the spatial position of the muon. This requirement is seen in an actual detection of muon decay, where the muon path and the path of the outgoing electron are explicitly shown (see Fig. 6.3). The original photographs are shown in [23] (p. 4) and in [67] (p. 26).

Figure 6.3 shows another aspect of muon decay. The decay process takes place at the tip of the angle of this figure. This means that *the experiment shows the spatial position of the decay.* QFT of the particles that are shown in this figure should explain this process; hence, this theory should have a coherent expression for density, which is the quantum notation of spatial position.

In principle, each outgoing particle is recorded by a specific detector. This detector records the energy-momentum and the detector's space-time point of the arriving particle. As usual, these data include an estimate of the measurement error. This information is used by experimenters to find out whether two requirements hold:

- The outgoing particles should emerge from a small space-time region.

- The process should conserve energy.

Experimenters calculate the common spatial region and the total invariant energy of the outgoing particles. If the results of these tests are satisfactory, then the event is recorded as muon decay.

These arguments explain why a theoretical description of the muon decay effect must yield coherent expressions for the measured quantities – the energy-momentum and the space-time position of the decaying particle of Fig. 6.2. The muon is a Dirac particle, and the Dirac theory provides the required expressions for its density and its energy-momentum. As a quantum theory, it abides by the Heisenberg uncertainty principle. Thus, we have a Dirac particle that moves in the vacuum. The four components of the spinor of a solution of a free spin-up Dirac particle are as follows (see [2], p. 30):

$$\psi = (1, 0, p_z/Q, p_+/Q)S \exp(i(\mathbf{k} \cdot \mathbf{x} - \omega t)), \qquad (6.3)$$

where

$$p_+ = p_x + ip_y, \quad Q = E + m, \quad S = \sqrt{Q/2m}. \qquad (6.4)$$

Here m and E denote the particle's mass and energy, respectively. As stated above, the density of a quantum particle is the 0-component of its 4-current. The required expression for the Dirac particle's 4-current is shown above (see (3.32) on p. 34). The explicit expression for the space-time position of a Dirac particle is (see e.g. [18], p. 24).

$$< \mathbf{r}(t) >= \int \psi_i^\dagger(\mathbf{r}, t)\mathbf{r}\psi_i(\mathbf{r}, t)d^3r. \qquad (6.5)$$

The foregoing points show how the quantum theory of a Dirac particle provides a coherent interpretation of the experiment. This issue is summarized as follows:

A. The Dirac equation is derived from the QED Lagrangian density.

B. The Dirac equation is solved for a free particle, and the solution ψ of (6.3) describes the particle's properties.

C. The Noether theorem is applied to the Dirac Lagrangian density and yields the Noether 4-current j^μ (3.32). The particle's density is the 0-component of the 4-current.

D. Subsection 4.4.2 on p. 52 explains how density is used for the construction of a Hilbert space of every Dirac particle and the operators that operate on the quantum functions of this space.

E. The mentioned subsection explains how the Hilbert spaces of all particles of Fig. 6.2 are utilized for the construction of the Fock space of these particles.

F. Weak interaction operators of the Fock space explain the muon decay process.

G. Position and momentum are conjugate variables and the present discussion shows an important aspect of the relations between them. Operators of density are applied to the quantum function ψ and provide an estimate of the particle's position. Other operators yield the energy-momentum of the decaying particle. This is the theoretical explanation of the elements of the muon decay experiment of Fig. 6.2. It means that these conjugate variables are equally important.

H. In particular, experimenters record the space-time position of the point where an outgoing particle hits its measuring device and the energy-momentum of this particle. This information enables them to determine a quite small space-time region from which all the outgoing particles emerged. Hence, a theory of a decaying particle must provide expressions for its space-time position and its energy-momentum. Relations (6.5) and (6.3) respectively satisfy these tasks.

I. Density appears in different places of the theoretical description of the decay process. It explains the position of the decaying process, as well as the position of the outgoing particles at the detectors. It is also used in the inner product of the particles' Hilbert spaces. The Fock space uses these Hilbert spaces and the particle's creation/destruction operators operate on entries of the Fock space.

J. It is interesting to note that the aforementioned arguments are independent of the specific QFT that describes the decaying process.

Conclusion: The Dirac theory of an elementary spin-1/2 particle explains the experimental attributes of the decay of such a particle. The muon decay is an example of this process.

Figures 6.2, 6.3 of p. 70 show how muon decay provides an example of the merits of the GCP in Fig. 3.2 (see p. 15). A description of the details of this issue is as follows: The incoming particle moves in the vacuum, and its motion is described by CPH. In particular, a solution of the classical equations of motion determines the particle's space-time position $r(t)$ and energy-momentum. Conservation of energy-momentum is a well-known effect, and it is utilized in a measurement of muon decay. The muon decay is explained by the destruction and creation operators of QFT. These operators operate on the Fock space, which is based on the particles' Hilbert spaces of QM. The outgoing particles move in the vacuum, and CPH explains their motion. QFT mediates between the initial and final states of muon decay. To explain this decay, it must provide coherent expressions for the density of the particles in Fig. 6.2 and their energy-momentum. This is what the GCP of Fig. 3.2 requires.

> *Conclusion: A QFT of a decaying particle must provide an expression for the time dependence of its density and its energy-momentum. This outcome makes sense because space-time and energy-momentum are conjugate variables.*

Chapter 7

Problems with Quantum Theories

This chapter discusses several topics that belong to quantum theories. These topics deserve special attention because many textbooks do not provide a comprehensive discussion of them.

7.1 No Problem with the Particle-Wave Duality

Subsection 4.3.2 on p. 47 shows that physical laws *prove* the principle of complementarity. Therefore, wave properties of a point-like particle are coherent theoretical elements, and complementarity is not an additional postulate that is required for the logical coherence of quantum theories.

7.2 The Need for an Interaction Term

An acceptable physical theory of a given system must provide a prediction that can be measured, and this issue boils down to an effect that can be sensed by humans. Generally, a change in the state of a physical system is not directly sensed by humans. Hence, a change in the state of a measuring device indicates a physical process where the system's state changes. In particular, a theory of an elementary particle must contain an interaction term that changes the particle's state. (It is pointed out above that the structure of general relativity is different. This theory uses the metric as the dynamical variable. However, as stated at the beginning of this book, gravitation is not discussed here.) Therefore, the QFT Lagrangian density of a given particle should contain an interaction term that depends on the particle's variables *and* on variables that represent an external field. Evidently, like all other terms of the Lagrangian density, this term must be a mathematically real expression that is a Lorentz scalar whose dimension is $[L^{-4}]$. To conclude, in QFT, the interaction term must be consistent with the theory of the examined particle/field *and* with the theory of the external field. It is shown later in this book that these self-evident requirements are ignored by several SM theories.

7.3 Problems with Mathematically Real Quantum Functions

It is shown in section 4.2 and subsection 4.4.1 that a mathematically real quantum function of a massive particle violates both the de Broglie principle and the correspondence with the Schroedinger theory. Relying on the correspondence principle of section 4.4, we can offer the following conclusion:

> In the case of a massive particle, a QFT of a mathematically real function is unacceptable.

This outcome holds for the real KG function, the electroweak description of the Z boson, the mathematically real Higgs boson, the Majorana neutrino theory, and the massive Proca photon. Other problematic points of these particles are mentioned later.

That SM textbooks do not show a coherent expression for the density of any mathematically real quantum function is another indication of the validity of this result. Unfortunately, SM textbooks too often refrain from stating that they do not show crucial expressions. This book makes a modest attempt to close this gap in SM textbooks.

7.4 Covaraince of the Lagrangian Density and the Hamiltonian Density

The discussion of section 3.5 shows that the Lagrangian density is a Lorentz scalar. In contrast, Eq. (3.21) shows that the Hamiltonian density is the T^{00} component of a second-rank tensor called the energy-momentum tensor.

> *Conclusion: A term of the Lagrangian density and its corresponding term of the Hamiltonian density undergo different Lorentz transformations.*

It is shown in this book that the Lagrangian-Hamiltonian covariance difference (LHCD) has far-reaching implications. Unfortunately, this self-evident issue has long gone unnoticed by SM textbooks. Below, chapter 11 proves that this omission explains why the electroweak theory has been created on an erroneous basis.

In the case of a Dirac particle, the LHCD may be observed from another point of view. As explained above, the Lagrangian density is a Lorentz scalar. The standard construction of an interaction term uses the product

$$\bar{\psi}\psi, \text{ where } \bar{\psi} \equiv \psi^\dagger \gamma^0, \tag{7.1}$$

which is a Lorentz scalar, and another Lorentz scalar that is a contraction of the Dirac γ^μ matrices with external fields (see, e.g., the QED Lagrangian density (3.33) on p. 35). In the case of the Hamiltonian, one notes that in CPH, interaction takes place at the particle's spatial location. Hence, the quantum Hamiltonian density must depend on the particle's density, which is the quantum description of spatial location

$$\rho = \psi^\dagger \psi. \tag{7.2}$$

It follows that the quantum Hamiltonian density requires that the Dirac function $\bar{\psi}$ of (7.1) should be replaced by Dirac function $\psi^\dagger$ of (7.2). Therefore, the Dirac matrix γ^0 of (7.1) should join the interaction term. This Dirac matrix is the 0-component of the 4-vector γ^μ. This means that the relativistic properties of a term of the Hamiltonian density and those of the same term of the Lagrangian density are different.

It is interesting to note that the Dirac function ψ is the generalized coordinate of the Dirac Lagrangian density, and $i\psi^\dagger$ is the momentum conjugate to this coordinate (see [14], p. 52). Hence, like in CPH, the Dirac Hamiltonian density is a function of the generalized coordinate ψ and its conjugate momentum $i\psi^\dagger$. This is another aspect that shows how the Dirac equation is embedded in the general structure of physics.

7.5   The Multi-Configuration Concept

The topic of multi-configuration is a quite strange episode in the progress of physics. The problem begins with the electronic structure of atoms that have more than one electron. Here some mathematical tricks have enabled the calculation of the ground state of the 2-electron He atom. However, it has been recognized that a different approach is needed for atoms with more than two electrons. The concept of configuration is useful for this problem. For example, one configuration of the ground state of the 3-electron Li atom is written as $1s^2 2s^1$. This notation means that two electrons are in the 1s closed shell and another electron is in the second s shell, denoted as 2s. The power-like numbers denote the number of electrons in each shell.

This subsection does not intend to provide a detailed description of the problem because a full textbook would be required to accomplish this. Hence, the text below only aims to convince readers that an accurate description of a multi-electron atomic state *cannot take the form of a single configuration.*

The laws of quantum theory determine the actual electronic state of an atom. The calculation is carried in the atomic rest frame, and the energy states of the system are derived as eigenfunctions of the Hamiltonian. Here the Heisenberg picture is used and the quantum functions are time-independent. The calculation must abide by conservation laws. Symmetries of the Hamiltonian require that the *total angular momentum J* and its projection J_z should be well defined. Since electrodynamics conserves parity, the parity of the state must also be well defined. The Pauli exclusion principle imposes another constraint: all electrons of a single configuration must be in an antisymmetric state.

The calculation may use spherical polar coordinates. If a single electron is bound by a force whose value is independent of the angular coordinates, then its state has a well-defined angular momentum. This is not true for atoms with more than one electron because each electron interacts not only with the nucleus but also with all other electrons. It turns out that the constraint of the total angular momentum *J* allows a single electron to be in an undetermined angular momentum.

To prove this assertion, let us examine the ground state of the Li atom. This is a stable state, and, as argued above, the time-independent Heisenberg picture is suitable for the construction of the Hilbert space. Here, each function of the Hilbert space has a specific configuration.

The state's angular momentum and parity are $J^\pi = 1/2^+$. This means that every acceptable configuration must have these quantum numbers. Consider, for example, the two-configuration function ψ that pertains to this state (not including the radial coordinates functions, which should be treated separately):

$$\psi = C_1 1s^2\, 2s^1 + C_2 1s^1\, 2s^1\, 3d^1 \qquad (7.3)$$

C_i are numerical coefficients. Because of the Pauli exclusion principle, the overall spin of the two $1s^2$ electrons of the first term is zero, and the $2s^1$ electron determines the total angular momentum of the state, $J = 1/2$. The parity is even because all the single-particle s states have even parity. Similar arguments apply to the second term. The d electron means that its spatial angular momentum is $\ell = 2$. Electrons of this term are in different shells, and there is no problem with the production of a spatial antisymmetric state. It means that their total spin can be coupled to a symmetric $S = 3/2$ state. Hence, the addition law of angular momentum proves that L, S of the state can be coupled to $J = 1/2$. The parity of the s and d shells is even. This means that the parity of the second term is also even. These arguments prove that the two terms of (7.3) are legitimate configurations of the Li atom ground state.

The Hamiltonian is a Hermitian operator and the two terms of (7.3) yield a 2×2 Hermitian matrix, which is a sub-matrix of the Hamiltonian matrix

$$H_{12} = \begin{pmatrix} a & c \\ c^* & b \end{pmatrix}. \qquad (7.4)$$

The off-diagonal matrix elements do not vanish for many pairs of legitimate terms. In this case, the eigenfunction is an appropriate sum of the terms of (7.3), which is obtained from diagonalization of the Hamiltonian's matrix. Let a' denote the lowest eigenvalue of H_{12}. An important law of Hermitian matrices proves that $a' < a$, $a' < b$. *This means that the addition of new configurations decreases the calculated ground state of the atom!* This short explanation proves the following assertion:

> *The electronic state of an atom of more than one electron (including its ground state) takes the form of a sum of configurations.*

This conclusion is not my invention[1]. It can be found in older textbooks (see [68], chapter XV; [69]), and there are published

---

[1] A personal remark: I've studied this issue in an undergraduate course on quantum mechanics, delivered by Yehuda Shadmi, a disciple of Joel Racah.

papers on this concept [70, 71]. For example, the first edition of the Condon and Shortley textbook [68] was published in 1935. It says that the off-diagonal matrix elements cause energy levels to "be pushed apart and results in an intermingling of character through linear combination of the ψ's of the interacting levels" (see p. 365). Furthermore, a relatively complicated mathematical machinery called *Wigner-Racah algebra* or *angular momentum algebra* has been constructed for facilitating the calculation [9, 72].

The off-diagonal elements of the Hamiltonian submatrix (7.4) contribute to the decrease of the lowest eigenvalue of the Hamiltonian. These elements represent quantities of the form $<\psi_j|H|\psi_i>$. Hence, this origin of the off-diagonal element of (7.4) is called a *configuration interaction* (CI) in the literature.

Textbooks explain the QM description of the atomic structure and its relevance to the Mendeleev periodic table. To this end, a comparatively simple, approximate approach to the problem has been devised. This is called *central-field approximation* (see e.g. [18], p. 277) or *self-consistent field method* (see e.g. [27], p. 232, [73], p. 612). These textbooks clearly state that this procedure is just an *approximation*! One result of this approach is that a multi-electron atomic state is described by a *single* configuration. For a reason that is still unclear, this approximation is now regarded as an accurate description of atomic states, taking the form of a single configuration. For example, Wikipedia – which represents the consensus – says [74]: "The configuration that corresponds to the lowest electronic energy is called the ground state. Any other configuration is an excited state." The Wikipedia item continues, "As an example, the ground state configuration of the sodium atom is $1s^2$, $2s^2$, $2p^6$, $3s^1$..."

The multi-configuration concept is not restricted to atomic physics. Indeed, like electrons, quarks are spin-1/2 particles, and the high energy of the proton constituents indicates that its state is described by a larger configuration mixture of quark states. It turns out that the erroneous single configuration concept is the underlying reason for the much ado about the (otherwise straightforward) effect called the *proton spin crisis* [75]. At present (August 2021) its Wikipedia item states, "The problem is considered one of the important unsolved problems in physics."

Conclusion: Unfortunately, the relatively old multi-configuration concept has fallen by the wayside. Nevertheless, this concept is also relevant to the understanding of the structure of strongly interacting systems, such as the proton. It provides a straightforward explanation for the effect that particle physicists call the proton spin crisis.

Chapter 8

Problems with Electrodynamics

This chapter shows examples proving that the accepted structure of electrodynamics contains erroneous elements.

8.1 QED Problems in the Literature

It is interesting to note that important physicists have already expressed serious qualms concerning the validity of the present QED structure. They refer to a fundamental QED process called renormalization. Dirac described it as a procedure of an "illogical character" [76]. He continued: "I am inclined to suspect that the renormalization theory is something that will not survive in the future, and that the remarkable agreement between its results and experiment should be looked on as a fluke." Feynman used a less formal terminology and stated that renormalization is "a dippy process" (see [77], p. 128). This approach is also mentioned in a textbook: "In the quantum theory, these divergences do not disappear; on the contrary, they appear to get worse, and despite the comparative success of renormalisation theory the feeling remains that there ought to be a more satisfactory way of doing things" (see [78], p. 390).

Before continuing with the analysis, there is a point we should make: The present textbooks use two different names for particles that are associated with electromagnetic fields – real photons and virtual photons (see [2], p. 111; [23], p. 11; [14], p. 5; [79], p. 5). Relying on this issue, one clearly sees an inherent problem with the

QED Lagrangian density (see (3.33) on p. 35). The different names for a real photon and a virtual photon mean that these entities must have *at least one physically meaningful different property*. Hence, the term of the QED Lagrangian density $-\frac{1}{16\pi}F_{\mu\nu}F^{\mu\nu}$ (see (3.33) on p. 35), which does not distinguish between these entities, must be wrong! Specific aspects of this argument are noted below.

8.2   Radiation Fields and Bound Fields

The standard form of Maxwell equations (3.24) has one inhomogeneous tensorial equation and one homogeneous tensorial equation. Let us examine the special case where all Maxwell equations are homogeneous

$$F^{\mu\nu}_{\ \ ,\nu} = 0, \quad F^{*\mu\nu}_{\ \ \ ,\nu} = 0. \tag{8.1}$$

Solutions of the homogeneous Maxwell equations (8.1) are called radiation fields (see [3], p. 116), and solutions of the inhomogeneous Maxwell equations (3.24) – where $j^{\mu} \neq 0$ at some space-time coordinates – are called bound fields. Let us put forward an assertion where the full consequences are presently (August 2021) ignored by the standard QED structure.

> *Assertion: Radiation fields and bound fields represent entirely different physical entities.*

It is explained below that this assertion is important. Several independent proofs substantiate it:

PR.1 Let us begin with an issue that is already known to readers of this book. The analysis of the energy-momentum tensor of electromagnetic fields is presented in section 5.4, p. 59. This analysis proves that radiation fields yield a coherent expression for this tensor. In contrast, there is a problem with the bound field. This result proves the assertion that bound fields and radiation fields are different physical entities.

However, a mathematical rule of thumb says that if an assertion can be proved in one way, it is likely that it can also be proved in different ways. The next points illustrate this issue.

PR.2 QM provides a good description of the states and processes of the hydrogen atom. These descriptions can be regarded as reliable experimental properties of this atom. Consider the ground state of the hydrogen atom and an incoming photon whose energy equals the difference between the states $2p$ and $1s$. This is an allowed transition (see [18], p. 264) and the hydrogen atom is excited to the $2p$ energy level. This process proves that the photon's angular momentum and parity are 1^-. (The photon is massless and its angular momentum takes only the two helicity components $s_z = \pm 1$.) This angular momentum and parity of the photon are documented in the Particle Data Group's annual report [29].

Let us examine the bound fields the hydrogen atom ground state. QM calculations of atomic spin and parity of this state are determined by the electronic state, and the electromagnetic bound fields are completely ignored (see, e.g., [2], pp. 52-60). This means that if bound fields represent a genuine physical particle, the spin-parity of such a particle are 0^+. In contrast, it is shown above that the photon's spin-parity are 1^-. The different spin of radiation fields and bound fields, as well as the Wigner constraint W.2 in section 3.7, prove the assertion that bound fields and radiation fields are different physical entities. Moreover, parity conservation of QED proves that radiation fields and bound fields are different physical entities.

PR.3 Consider the Lorentz invariants of the electromagnetic fields

$$Inv_1. = B^2 - E^2 \tag{8.2}$$

and

$$Inv_2. = \boldsymbol{E} \cdot \boldsymbol{B} \tag{8.3}$$

(see [3], p. 68). The invariants (8.2) and (8.3) vanish for radiation fields emitted from a specific system (these relations are obtained from Eq. (66.8) of [3], p. 186 and from Eqs. (9.4) and (9.5) of [28], p. 392). In contrast, (8.2) does not vanish for the bound field of a single charge; this can be seen at its instantaneous rest frame, where retardation arguments prove that $\boldsymbol{E} \neq 0$ and $|\boldsymbol{B}| \ll |\boldsymbol{E}|$ in the region close to the charge.

> Conclusion: Relativistic attributes prove the assertion that bound fields and radiation fields are different physical entities.

PR.4 Let us examine another kind of electromagnetic interaction – a scattering process where the mediating fields are bound

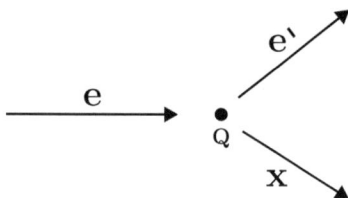

Figure 8.1: *Electron e is scattered by a charged particle Q.*

fields of two colliding particles (see fig. 8.1). An electron e moves from left to right and is scattered by an electrically charged particle Q. Here, e' denotes the 4-momentum of the outgoing electron and X denotes the sum of the 4-momentum of all other outgoing particles.

The 4-momentum transfer q^μ is the difference between the 4-momentum of the incoming electron and that of the outgoing electron, where q^μ is a space-like 4-vector (see [23], p. 190). In contrast, constraint W.3 of section 3.7 means that a genuine physical particle cannot have a space-like 4-momentum. It follows that in a scattering process, no genuine physical particle is exchanged between the colliding particle. This is another example showing the inherent difference between a real photon and bound fields. Furthermore, this inherent difference between interaction mediating fields and a genuine particle applies to all cases that are described by the scattering experiment of Fig. 8.1. In the case of electromagnetic fields, this difference is (at least implicitly) recognized by the general community: Physical photons are called *real photons*, whereas the term *virtual photons* is used to describe energy and momentum associated with an interaction of bound electromagnetic fields (see e.g. [2], p. 111).

This is another proof of the assertion that bound fields and radiation fields are different physical entities.

PR.5 Let us examine a motionless electron located at an external field-free region. Fundamental physical principles say that the mass/energy of this system is the electron's mass m. In contrast, the electron is an electrically charged particle, and its electric field does not vanish. This field is bound to the electron. It is well known that the Dirac equation (3.31) adequately describes this system. For convenience, let us rewrite this equation:

$$i\frac{\partial \psi}{\partial t} = [\boldsymbol{\alpha} \cdot (-i\boldsymbol{\nabla} - e\boldsymbol{A}) + \beta m + e\Phi]\psi. \qquad (3.31)$$

The Dirac equation (3.31) means that the 4-potential of the electron's field makes a null contribution to the mass/energy of this single-electron system. In contrast, the energy of radiation fields is positive. This is another example of the inherent difference between radiation fields and bound fields.

PR.6 Another contradiction arises if the momentum density of the electron's bound field is treated according to the Poynting 3-vector

$$\boldsymbol{S} = \boldsymbol{E} \times \boldsymbol{B}/4\pi \qquad (8.4)$$

(see [3], p. 81; [28], p. 237). The integral of a Lorentz transformation of this momentum density of the electron's bound field yields an incorrect value that is 4/3 times greater than the required value (see e.g. [80], chapter 28). This incorrect value is another proof of the inherent difference between bound fields and radiation fields.

As found in section 4.3.1, CPH and quantum theory say that an elementary particle like the electron is point-like. The null energy of its bound field relieves theoretical physics from the infinite self-energy problem of the field of a point-like particle and from the 4/3 problem of the Lorentz transformation of the electromagnetic momentum.

Each of the arguments made above proves that bound fields and radiation fields are different physical entities.

The structure of the QED Lagrangian density (3.33) is inconsistent with the assertion proved herein. Indeed, (3.33) has a term that is a product of the electromagnetic tensor $F^{\mu\nu}$ with itself. This tensor is a sum of radiation fields and bound fields $F^{\mu\nu} = F_R^{\mu\nu} + F_B^{\mu\nu}$, where the subscripts R and B denote radiation fields and bound fields, respectively. Introducing this sum into the product of the fields of the QED Lagrangian density (3.33), one finds

$$F_{\mu\nu}F^{\mu\nu} = F_{R\,\mu\nu}F_R^{\mu\nu} + F_{B\,\mu\nu}F_B^{\mu\nu} + 2F_{B\,\mu\nu}F_R^{\mu\nu}. \qquad (8.5)$$

(For the simplicity of the notation, the numerical factor $-1/16\pi$ is removed from (8.5).) Item PR.2 proves that if bound fields represent a particle, the particle's parity is even. Hence, the parity of the last term of (8.5) is odd. This term is unacceptable for the Lagrangian density of a parity conserving theory like QED.

8.3   Problems with the Electromagnetic 4-potential

MLE is independent of the electromagnetic 4-potential $A_\mu(t, \boldsymbol{x})$. This means that any mathematically correct manipulation of this quantity is acceptable in MLE. Hence, in MLE, the 4-potential can be regarded as an auxiliary mathematical quantity. This section contains an analysis of the 4-potential in VE, and classical and quantum aspects of this topic are examined. The analysis proves erroneous points of QED.

8.3.1   The Strange Status of the Electromagnetic 4-potential

Some may be surprised that well-known contemporary textbooks make contradictory statements about the relativistic properties of the electromagnetic 4-potential $A^\mu(t, \boldsymbol{x})$. Here, four quotations demonstrate this strange situation; two say that $A^\mu(t, \boldsymbol{x})$ is a 4-vector, whereas the other two say the opposite:

A.1 Landau and Lifshitz say: "Thus the action function of a charge in an electromagnetic field has the form

$$ S = \int_a^b \left(-mc\, ds - \frac{e}{c} A_\mu dx^\mu \right). \qquad (\text{``}16.1\text{''}) $$

The three space components of the 4-vector A^μ form a three-dimensional vector $\boldsymbol{A}$ called the *vector potential* of the field. The time component is called the *scalar potential*; we denote it by $A^0 = \Phi$. Thus

$$ A^\alpha = (\Phi, \boldsymbol{A})'' \qquad (\text{``}16.2\text{''}) $$

(see [3], p. 48, where the original equation numbers are shown here).

A.2 Feynman makes the following statement: "In short,

$$ A_\mu = (\Phi, \boldsymbol{A}) $$

is a four-vector. What we call the scalar and vector potentials are really different aspects of the same physical thing. They belong together. And if they are kept together the relativistic invariance of the world is obvious. We call A_μ the *four-potential*" (see [80], chapter 25).

A.3 By contrast, Weinberg's textbook examines the 4-potential of a photon and explicitly states: "The fact that A^0 vanishes in all Lorentz frames shows vividly that A^μ cannot be a four-vector" (see [20], p. 251).

A.4 Similarly, the textbook of Bjorken and Drell analyzes the 4-potential of electromagnetic fields and states that "we lose manifest Lorentz and gauge covariance" (see [26], p. 73).

The textbooks (see [20] p. 251, [26], p. 89) show how the gauge transformations of the 4-potential can be used for amending the problem. Considering this state of affairs, let us examine the definition of a 4-vector: "In general, a set of four quantities A^0, A^1, A^2, A^3 which transform like the components of the radius four-vector x^μ under transformation of the four-dimensional coordinate system is called a *four-dimensional vector (four-vector)* A^μ" (see [3], p. 15). This means that the QED 4-potential *is not a 4-vector*. To correct this issue, textbooks use a gauge transformation. In other words, the above-mentioned 4-vector definition and the application of gauge transformations means that textbooks argue that one can proceed *even though the 4-potential $A_\mu(t, \boldsymbol{x})$ is not a 4-vector.*

The foregoing quotations demonstrate that well-known textbooks make inconsistent statements concerning the problem of whether the electromagnetic 4-potential $A_\mu(t, \boldsymbol{x})$ is a genuine 4-vector. This state of affairs highlights the need for an adequate analysis of this issue, and the rest of this section is dedicated to this objective.

8.3.2 The 4-potential of a Charged Pointlike Particle and its Electromagnetic Fields

Consider a charged elementary point-like classical particle e and a space-time point $x = (t, \boldsymbol{x})$. Its 4-potential is called the Lienard-Wiechert 4-potential, and its values at x are expressed as (see [3], p. 174; [28], p. 656)

$$A_\mu = e \frac{v_\mu}{R^\alpha v_\alpha}. \tag{8.6}$$

Here R^μ denotes the 4-vector from the retarded position of the charge to the measurement point x, and v^μ denotes the charge's retarded velocity. This 4-potential yields the following electromagnetic fields (see [3], p. 175; [28], p. 657)

$$\boldsymbol{E} = e \frac{1 - v^2}{(R - \boldsymbol{R} \cdot \boldsymbol{v})^3} (\boldsymbol{R} - \boldsymbol{v}R) + \frac{e}{(R - \boldsymbol{R} \cdot \boldsymbol{v})^3} \boldsymbol{R} \times [(\boldsymbol{R} - \boldsymbol{v}R) \times \boldsymbol{a}] \tag{8.7}$$

and

$$\boldsymbol{B} = \boldsymbol{R} \times \boldsymbol{E}/R; \qquad (8.8)$$

$\boldsymbol{a}$ denotes the retarded 3-acceleration.

At spatial regions that are far from the retarded position of the charge, the first term of (8.7) and (8.8) decreases like $1/R^2$. The fields' energy density and the Poynting vector depend on the product of the fields (see (5.15) and (5.16) on p. 61). Hence, a field that decreases like $1/R^2$ represents bound fields. This term is acceleration-free, and in the literature, it is called a *velocity field*. In contrast, in these regions, the second term of (8.7) and (8.8) decreases like $1/R$. Hence, this term pertains to radiation fields (see [3], pp. 175, 176; [28], p. 657). This term depends on acceleration, and in the literature, it is called an *acceleration field*.

The analysis of section 8.2 proves that velocity fields and accelerations fields represent entirely different physical entities. This means that the Lienard-Wiechert 4-potential (8.6) yields two entirely different kinds of fields. This result casts serious doubts concerning the connection between the Lienard-Wiechert 4-potential (8.6) and physical reality. This outcome encourages a further analysis of this topic.

8.3.3   The Multiparticle Aspect of Radiation Fields

Consider the two radiating systems of Fig. 8.2. Panel (A) of the figure shows a single charge q that moves uniformly along a circle, which is embedded in the (x, y) plane, and its center coincides with the origin of the coordinates. The point p lies on the z-axis

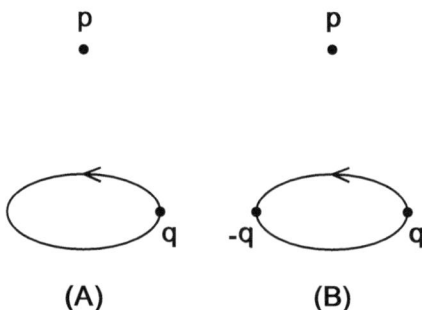

Figure 8.2: Two radiating systems. (A) Charge q moves uniformly along a circle. (B) Two charges, $\pm q$, move uniformly along a circle.

at the radiation zone. The circular motion of charge q proves that it accelerates towards the circle's center. This acceleration $\boldsymbol{a}$ and formulas (8.7), (8.8) indicate that the radiation fields do not vanish at point p.

Let us examine the radiation fields $\boldsymbol{E}$ and $\boldsymbol{B}$ at point p and compare the fields of panel (A) in Fig. 8.2 with those of panel (B). In panel (B), there are two charges $\pm q$ located at two antipodal points of the circle. These charges move along the circle with the same velocity as that of charge q in panel (A). For point p, the retarded time of charge $+q$ is the same as that of charge $-q$. Formulas (8.7), (8.8) prove that, at point p, the electric radiation field and the magnetic radiation field of charge $+q$ are the same as those of charge $-q$. This means that at point p, the radiation fields $\boldsymbol{E}$, $\boldsymbol{B}$ of panel (B) are twice as strong as those of panel (A). Now, the Poynting vector (8.4) determines the energy current of electromagnetic fields.

This vector proves that *at point p, the energy current of panel (B) of Fig. 8.2 is four times greater than that of panel (A) of this figure.* Furthermore, since the same frequency holds for the fields of the two cases, one can conclude that at the vicinity of point *p, the number of photons of panel (B) of the figure is four times greater than the number of photons of panel (A).* This means that the actual radiation emitted by a system of charges is *not* the sum of radiation emitted by individual charges. This example proves the important conclusion defined below.

> *The radiation emitted from a system of charges is a multi-charge effect. In particular, a photon cannot be assigned to a specific charge of the radiating system.*

Realizing this conclusion, let us compare the 4-potential of the QED Lagrangian density (3.33) of the local system with the Lienard-Wiechert 4-potential (8.6) of several charges of the radiating system. The 4-potential A_μ of the QED Lagrangian density (3.33) pertains to the electromagnetic interaction of the charged particle that is represented by the function $\psi(t, \boldsymbol{x})$, where $(t, \boldsymbol{x})$ denote the space-time location of the charge described by this quantum function. Hence, the QED Lagrangian density (3.33) requires an expression for the 4-potential $A_\mu(t, \boldsymbol{x})$ that is a function of the same local coordinates as those of $\psi(t, \boldsymbol{x})$.

An observation of the Lienard-Wiechert 4-potential (8.6) clearly proves that it is unsuitable for a description of incoming radiation of the QED Lagrangian density (3.33). The reason is that (8.6) is a function of local coordinates *and* retarded coordinates of the charges of the radiating system. In particular, R^μ of this 4-potential depends on local coordinates and retarded coordinates. The velocity v^μ is the retarded velocity of a charge at the radiating system. On this basis, one arrives at the conclusion below.

> *Conclusion: In the case of incoming radiation, the single-particle Lienard-Wiechert 4-potential (8.6) is unsuitable as a part of the interaction term of the QED Lagrangian density.*

8.3.4   A Relativistically Consistent 4-potential of Radiation Fields

The discussion in section 3.5 shows that the dimension of the Lagrangian density is $[L^{-4}]$. The term of the QED Lagrangian density (3.33) that contains the 4-potential A_μ also contains the product $\bar\psi\psi$ of the Dirac function, and the dimension of this product is $[L^{-3}]$. Hence, the electromagnetic interaction factor of the QED Lagrangian density must be a quantity with the dimension of $[L^{-1}]$. This is the dimension of the 4-potential A_μ. Moreover, the Dirac matrices γ^μ are entries of a 4-vector, and in (3.33), this 4-vector is contracted with A_μ. Hence, A_μ must be a 4-vector. However, as proved above, if one demands that A_μ should be a function of the local coordinates $(t,\boldsymbol{x})$ of the quantum function $\psi(t,\boldsymbol{x})$, $A_\mu(t,\boldsymbol{x})$ is not a 4-vector.

This outcome appears to be ominous property that threatens the consistency of VE in general and QED in particular. A prescription for how to overcome this problem is described here. Let us formulate the problem: SR shows how a 4-vector of a relativistically coherent theory transforms under a Lorentz transformation. Here, we have a different issue.

> *Problem: What is the relativistically coherent transformation of the 4-potential $A_\mu(t,\boldsymbol{x})$, which is not a 4-vector?*

Below, this problem is called the *4-potential problem*.

Section 8.2 proves that radiation fields and bound fields are different physical entities. Here, the 4-potential problem is solved for radiation fields. (For bound fields, see the open problem U.F on p. 237.) Item PR.3 of section 8.2 proves that the two invariants of the electromagnetic fields vanish for radiation fields that are emitted from a given source. The electromagnetic fields solve the Maxwell equations, meaning that they are coherently defined at every space-time coordinate $(t, \boldsymbol{x})$. Therefore, in every inertial frame, the electromagnetic radiation fields $\boldsymbol{E}$, $\boldsymbol{B}$ are perpendicular to each other and $|\boldsymbol{E}| = |\boldsymbol{B}|$. These properties of the electromagnetic fields enable the construction of a relativistically coherent 4-potential in every inertial frame. For radiation wave of frequency ω that moves in the z-direction and it is polarized in the x-direction, the 4-potential is

$$A_\mu = (0, A\exp(i(kz - \omega t)), 0, 0), \qquad (8.9)$$

where A is the wave's amplitude and $k = \omega$ determine the undulating part of the electromagnetic wave (see also [81]). (The generalization of this proof to any radiation fields is straightforward.)

> *Conclusion: Eq. (8.9) is a relativistically coherent solution for the 4-potential problem.*

It has been proved in this section that the 4-potential is *not* a 4-vector. Nevertheless, it can be used coherently for electromagnetic radiation. As in MLE, the 4-potential can be used in VE as an auxiliary mathematical expression if it is restricted to radiation fields.

8.4 Problems with the Gauge Transformation

In MLE, a gauge function $\Lambda(x)$ changes the 4-potential A_μ

$$A_\mu(x) \rightarrow A_\mu(x) + \Lambda(x),_\mu. \qquad (8.10)$$

Here the gauge function $\Lambda(x)$ is an arbitrary function of the four space-time coordinates. The gauge transformation (8.10) does not alter the electromagnetic fields:

$$F'_{\mu\nu} = A_{\nu,\mu} - A_{\mu,\nu} - \Lambda(x),_{\nu,\mu} + \Lambda(x),_{\mu,\nu} = F_{\mu\nu}, \qquad (8.11)$$

where $F'_{\mu\nu}$ denotes the gauge-transformed electromagnetic field. As explained in subsection 3.6.1 on p. 29, MLE is independent of the 4-potential A_μ. Hence, MLE is a gauge-invariant theory, and a gauge transformation is a legitimate operation within this theory. This means that in MLE, a gauge transformation can be regarded as an auxiliary mathematical operation.

As stated above, MLE and VE are not identical theories. It is shown here that the gauge issue is an example that substantiates this claim. In VE, the 4-potential is used explicitly in the calculation of the action. For example, the actual form of the QED Lagrangian density (3.33) depends explicitly on the fields $F^{\mu\nu}$ *and* on the 4-potential A_μ. Therefore, unlike in the case of MLE, the issue of whether VE is a gauge-invariant theory requires further examination. This section discusses the consequences of the introduction of gauge transformations into the variational principle, and VE is examined below.

Unlike the case of MLE and its gauge transformation (8.10), in QED, a gauge transformation changes the 4-potential *and* the phase of the quantum function of a Dirac particle (see [14], p. 78; [20], p. 345):

$$A_\mu(x) \to A_\mu(x) + \Lambda(x)_{,\mu}, \qquad \psi(x) \to \exp(-ie\Lambda(x))\psi(x), \quad (8.12)$$

where the symbol e in the exponent denotes the electronic charge. As stated above, a gauge transformation uses a gauge function $\Lambda(x)$, which is an *arbitrary* function of the four space-time coordinates $x \equiv (t, \boldsymbol{x})$.

Before entering into details, one should recognize that a gauge transformation aims to add a 4-vector to the potential A_μ. However, it is proved above that the electromagnetic 4-potential is *not* a 4-vector. Hence, a gauge transformation cannot be an element of a physically valid theory. More details are discussed below.

A straightforward substitution of the gauge transformation (8.12) into the QED Lagrangian density (3.33) proves that the QED Lagrangian density is invariant under this gauge transformation. Indeed, $\bar{\psi}\psi$ is a mathematically real quantity that is independent of the phase. Furthermore, the change of the 4-potential A_μ is canceled out by the corresponding quantity obtained from the partial derivative $i\partial_\mu$ of (3.33).

The consistency of the gauge transformation (8.12) is analyzed below, relying on Theorem A.

> *Theorem A: If a Lagrangian function is invariant under a certain transformation, the entire theory that is derived from this Lagrangian function is invariant under that transformation provided the Lagrangian function and the transformation are free of mathematical contradictions.*

Gauge transformations play a key role in the SM structure. For example, at present (August 2021), the gauge item is described as follows on Wikipedia: "Gauge theories are important as the successful field theories explaining the dynamics of elementary particles." Similarly, a textbook states (see [14], p. 78), "A crucial property of the QED Lagrangian is that it is invariant under the gauge transformation" (the gauge transformation (8.12) follows this quotation). Hence, to deny gauge transformation, it is better to show multiple arguments where each disproves this concept. The following points show several fundamental errors of the QED gauge transformation (8.12):

G.1 The power series expansion of the exponential function of the gauge transformation (8.12) is

$$\exp(-ie\Lambda(x)) = 1 - ie\Lambda(x) + ... \qquad (8.13)$$

A fundamental law of physics says that all terms of a physically valid expression must have the same dimension. Furthermore, in the case of a relativistic expression, these terms must undergo the same Lorentz transformation. The first term on the right-hand side of (8.13) is the pure number 1, which is a dimensionless Lorentz scalar. The same is true for the imaginary number i, and in the units used herein, the electric charge e is also a dimensionless Lorentz scalar. It follows that the gauge function $\Lambda(x)$ must be a dimensionless Lorentz scalar. This constraint is violated by the intrinsic arbitrariness of the gauge function used by gauge theories. Moreover, the homogeneity of space-time proves that any function of the coordinates that is a dimensionless Lorentz scalar must be a numerical constant. Hence, a gauge function that is an arbitrary function of the space-time coordinate cannot be used in the exponential function of the gauge transformation (8.12).

G.2 The de Broglie principle defines the relation between the wavelength of a quantum particle and its momentum (see section 3.5)

$$\mathbf{k} = \mathbf{p}. \qquad (8.14)$$

It follows that an application of the quantum momentum operator $\mathbf{p} = -i\nabla$ to the exponential factor of (8.12) proves that to abide by the momentum value of the particle, the gauge function must be independent of the spatial coordinates. This result is inconsistent with the intrinsic arbitrariness of the gauge function used by gauge theories.

G.3 The significance of the Hamiltonian is noted in several places of this book (see, e.g., subsection 3.2.1 and section 3.5). The Dirac Hamiltonian density is derived from the QED Lagrangian density (3.33). Applying (3.21) to the Dirac function of (3.33), one can find the Dirac Hamiltonian density

$$\mathcal{H}_D = \pi \dot{\psi} - \mathcal{L} = \psi^\dagger [\boldsymbol{\alpha} \cdot (-i\boldsymbol{\nabla} - e\boldsymbol{A}) + \beta m + e\Phi]\psi, \quad (8.15)$$

where $\boldsymbol{\alpha}$ and β are the ordinary Dirac matrices, and the relation $\bar{\psi} = \psi^\dagger \gamma^0$ is used. The product $\psi^\dagger \psi$ is the density of the Dirac particle. Hence, the Dirac Hamiltonian is

$$H_D = \boldsymbol{\alpha} \cdot (-i\boldsymbol{\nabla} - e\boldsymbol{A}) + \beta m + e\Phi. \quad (8.16)$$

A gauge transformation changes the 4-potential $(\Phi, \boldsymbol{A})$. However, contrary to the QED Lagrangian density (3.33), the Dirac Hamiltonian (8.16) is independent of the time derivative of the Dirac function. Hence, an arbitrary gauge transformation may change the scalar potential Φ. This change of the Dirac Hamiltonian is not canceled by an appropriate time derivative. The Hamiltonian is the energy operator. Therefore, one arrives at the conclusion below.

> Conclusion: The QED gauge transformation violates energy conservation.

Here is an example that illustrates this conclusion. Let us see what happens in the simple case of a motionless Dirac particle which is in a field-free region where $A_\mu = 0$. In this case, the Dirac equation is

$$i\frac{\partial \psi}{\partial t} = H_D \psi = \boldsymbol{\alpha} \cdot \boldsymbol{P} + \beta m. \quad (8.17)$$

It follows that the energy of the system is $E = m$. Now, let us introduce the gauge function

$$\Lambda(t, \boldsymbol{x}) = t^2, \quad (8.18)$$

which is a legitimate example of an arbitrary function of the four space-time coordinates. The new 4-potential that is derived from the gauge function (8.18) is

$$A_\mu = (2t, 0, 0, 0). \quad (8.19)$$

Substituting (8.19) into the Dirac Hamiltonian (8.16), the energy of the system in this case becomes

$$E = m + 2et. \tag{8.20}$$

Result (8.20) demonstrates an unacceptable contradiction: It says that the energy of a motionless particle in a field-free space-time region increases with the increase of the time t. This means that a legitimate gauge transformation violates the law of energy conservation. Another error of (8.20) is the violation of the requirement of a uniform dimension of all terms of a physical expression. Indeed, the dimension of mass is $[L^{-1}]$, and the dimension of time is $[L]$. The charge e is dimensionless.

G.4 An important element of the gauge transformation (8.12) is the change of the quantum particle's phase. Therefore, one kind of test of gauge transformation compatibility is an examination of an interference experiment where the particle's phase plays a crucial role. Here, the calculations adhere to the inherent gauge transformation property where the gauge function is an *arbitrary* function of the four space-time coordinates.

Consider an electron that moves parallel to the z-axis. The external fields vanish and the electron moves inertially. The components of the electron's momentum are

$$k_x = k_y = 0, \; k_z = 1. \tag{8.21}$$

The electron's wave function takes the form

$$\psi(x) = \exp[i(z - \omega t)]\chi, \tag{8.22}$$

where ω and χ denote the frequency and a Dirac spinor, respectively.

The electron passes through two slits and the two sub-beams interfere on screen S (see fig. 8.3). The sine curve of this figure represents the intensity of the constructive/destructive interference at the corresponding points on S. This curve is obtained from a straightforward calculation of the phase of the two sub-beams. Let us examine how the gauge transformation

$$\Lambda(x) = -z/e \tag{8.23}$$

affects the interference calculation. Here, the gauge function depends only on the z-coordinate. Using the specific value

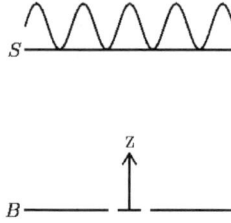

Figure 8.3: *Electronic beam moving parallel to the z-axis. The beam hits the barrier B and two sub-beams pass through the slits. Later, the sub-beams interfere on screen S.*

$k_z = 1$, one finds that substitution of the gauge transformation (8.12) and the gauge function (8.22) into the electron's function (8.22) yields the following wave function

$$\psi(x) = \exp(-i\omega t)\chi. \qquad (8.24)$$

Evidently, the z-dependence of the gauge function (8.23) cancels the z-dependence of the original electron's phase of (8.22). The outcome of the gauge transformation of (8.23) is a wave that has a uniform spatial distribution. It follows that there is no constructive/destructive interference, and the interference pattern of Fig. 8.3 is destroyed.

The interference pattern of Fig. 8.3 represents the real experimental data. It follows that the introduction of a gauge transformation into QED destroys its compatibility with experiments.

This section shows several independent arguments, each of which disproves the acceptability of gauge transformations in QED. Theorem A of this section and the errors of the gauge transformations disprove the merits of the invariance of the QED Lagrangian density with respect to gauge transformations. This result and the crucial role of gauge transformations in the SM demonstrate inherent QED errors.

8.5   Consequences of QED Problems

The above-mentioned inherent QED problems affect some physical theories that rely on an erroneous QED element. The primary result of this chapter means that it is high time for a thorough reexamination of QED. This assignment can certainly do no genuine scientific damage. Several specific cases are discussed below.

8.6 Problematic Results of the 4-Potential – the Dirac Monopole Theory

The Dirac monopole theory [4,33] derives the electromagnetic fields of a charge and a monopole from the same 4-potential. This is a dubious approach because it is proved here that the 4-potential is not a 4-vector, and it can be used only as an auxiliary mathematical expression. Indeed, quite a few erroneous elements of the Dirac monopole theory are already discussed in subsection 3.6.2.

8.7 Problematic Results of the 4-Potential – the Electroweak Theory

The electroweak theory regards the Z particle as an elementary particle that is described by a 4-vector. This 4-vector has close relationships with the electromagnetic 4-potential (see [46], p. 307). However, this point is doubtful because the electromagnetic 4-potential is not a 4-vector, as shown above. Indeed, it is proved later in this book that the electroweak theory contains many inherent errors (see chapter 11).

8.8 Problematic Results of the 4-Potential – the Aharonov-Bohm Effects

Aharonov and Bohm (AB) published articles that insist on the significance of the electromagnetic 4-potential as a physically meaningful quantity [82, 83]. However, their idea certainly cannot be correct for several reasons:

Pot.1 It is proved in this chapter that the 4-potential is not a genuine 4-vector. Hence, it cannot be a fundamental element of a coherent physical theory. Thus, the 4-potential is just an auxiliary quantity that may be used for facilitating calculations.

Pot.2 It is proved in this chapter that radiation fields and bound fields are different physical objects. The Lienard-Wiechert 4-potential (see subsection 8.3.2 on p. 89) is a quantity that yields these two completely different physical objects. Hence, it cannot be a physically coherent quantity, and like other 4-potential expressions, it is just an auxiliary quantity.

Pot.3 On top of the previous point, the Darwin Lagrangian includes the next term of the velocity-dependent expansion of the two-charge interaction in terms of a mechanical-like instantaneous two-body interaction (see [3], p. 179). This Lagrangian removes not only the 4-potential but also the electromagnetic bound fields. The Breit interaction is the quantum version of the interaction term of the Darwin Lagrangian. It uses the Dirac velocity operator $\boldsymbol{\alpha}$ instead of the Darwin classical velocity $\boldsymbol{v}$. As a power series expansion in terms of velocity, the Darwin-Breit interaction certainly applies to a low-velocity system. It turns out that AB also examine this kind of system. For example, they state in [82], "In the nonrelativistic limit (and we shall assume this almost everywhere in the following discussions)...". In the second AB article [83], the authors speak about adiabatic changes in the potential, which represent a very slow process. Hence, in these cases, the Darwin-Breit interaction removes not only the 4-potential but also the electromagnetic bound fields, and the entire AB assertion collapses [83].

The AB idea is more than 60 years old, and the mainstream literature contains many supporting articles that discuss its details. Therefore, it is important to substantiate the previous general remarks on the 4-potential with specific arguments that examine specific AB claims.

The first AB article [82] discusses the phase of a *single* electron and its interactions. This is incorrect because the phase is proportional to the action of the system, and each term of a quantum state describes all quantum elements of the system. Each term has one common phase for these elements. In their second paper [83], AB claim that their arguments hold for the entire system.

The AB idea is divided into two effects, called the electric AB effect and the magnetic AB effect. Below, these effects are discussed separately.

8.8.1   The Electric AB Effect

Figure 8.4 demonstrates the electric AB effect [82]. (Because of page width limits, the x-axis of the figure is contracted significantly.) An electronic beam moves from left to right, and at a

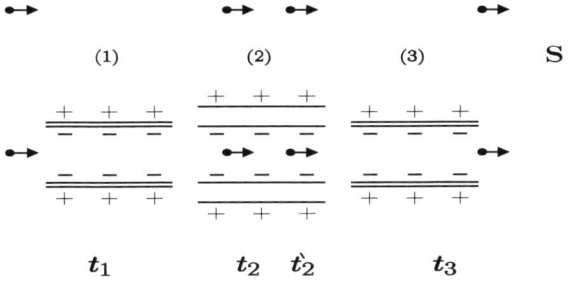

Figure 8.4: *The electric AB effect (see text).*

point outside the figure, it is chopped into well-separated portions. Later, at another point (still unseen in the figure), a beam splitter splits it into two sub-beams that continue to move rightward. A black spot describes the position of an electron of each sub-beam at a specific instant, and the arrow shows the direction of its motion. The apparatus comprises an upper pair and a lower pair of large thin plates made of insulating material. Each plate is covered uniformly with positive or negative charge of the same density. Hence, each plate is an inert object because its state does not change, and all its elements can only move as a rigid body. The state of the apparatus changes during the experiment, and the figure shows the position of its plates at three time intervals. At $t, t < t_1$, the distance between the two plates of each pair is infinitesimal. Considering the plates' opposite charge, one finds that the plates' electric potential and their field cancel each other at the electron's position in this period.

Let us examine the state of an electron of the lower sub-beam at different instants. At t_2, the electron is well inside the plates. Around this time, an engine of the apparatus drives outwards the positively charged plates, and they reach the position shown in the middle part of the figure. The electron is far from the plates' edges, and it continues its motion in a field-free region. However, the electric potential is negative at the electron's location. At t_2', the electron is still well inside the plates, and the positively charged plates return to their original position. At t_3, the electron exits the plates' region and continues its motion towards the interference screen S.

The arguments above show that the electron of the lower sub-beam travels in a field-free region. Its potential at different instants was as follows: For $t, t < t_2$, it moved in a null potential; for

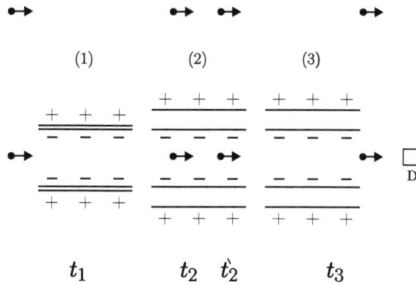

Figure 8.5: *Refutation of the electric AB effect.*

t, $t_2 < t < t'_2$, it moved in a negative potential; and for t, $t > t'_2$, it moved in a null potential. During this time, an electron of the upper sub-beam moved in a null potential and field-free region. Electrons of the two sub-beams interfere on screen S.

AB argued that the potential variation affects the phase of an electron of the lower sub-beam. As a result, the interference pattern changes. In their view, this outcome proves the physical meaning of the electric potential.

Claim No-AB1: The electric AB assertion is wrong because it violates energy conservation.

Proof: The following argument disproves the AB assertion: Let us examine a modification of the previous experiment (see fig. 8.5). Until t'_2, the motion is the same as that of fig. 8.4. In the second experiment at t, $t > t'_2$, the state of the plates remains unchanged until the electron exits the plates' region. Because of the negative potential between the two pairs of plates, the negatively charged electron of the lower sub-beam increases its kinetic energy. It enters a device D that removes the additional kinetic energy, and the electron restores its original kinetic energy. The screen S of the previous experiment is removed. For each sub-beam, appropriate magnetic fields change the direction of the electron's motion, unify the two sub-beams, and bring them back to the left-hand side of the apparatus, where the unified beam enters the beam chopper. During the latter period, the positively charged plates return to their original position.

If AB are right, then the net outcome of this process is a gain of energy by device D, which is a sheer violation of the energy conservation law. This totally unphysical result proves Claim No-AB1. It is shown later that the foregoing discussion not only disproves the theoretical basis of the electric AB effect but also shows where the original AB flaw rests.

In their second paper, AB claimed to extend the argument of the first paper, stating that they "include the sources of potentials quantum-mechanically, and we show that when this is done, the same results are obtained as those of our first paper in which the potential was taken to be a specified function of space and time." However, the previous example proves that this AB assertion is not true. Another argument substantiates this claim. Let us express the single-particle state of an electron of the split sub-beam as

$$\psi = a\psi_o(x) + b\psi_i(x), \qquad (8.25)$$

where the subscripts o and i denote an electron that is outside or inside the plates' region, respectively, and a and b are numerical constants. Let Ψ denote the quantum state of the apparatus. Evidently, the net effect of $\psi_o(x)$ on the plate's motion vanishes. In contrast, the electron's negative charge means that $\psi_i(x)$ resists the outward motion of each of the two positively charged plates. Hence, for the state $\psi_i(x)$, the plates' self-energy is a bit smaller than that of the case described by $\psi_o(x)$. The combined QM state is

$$\Phi = a\psi_o(x)\Psi_o(y_j, t) + b\psi_i(x)\Psi_i(y_j, t), \qquad (8.26)$$

where y_j denote the coordinates of elements of the apparatus.

Equation (8.26) agrees with one term of the sum of Eq. (19) of the second AB paper [83]. Below their Eq. (19), AB commented: "... because the parts of the source are so heavy, only a single such product is actually needed." The experiment described above demonstrates the incorrectness of this AB statement. Evidently, at $t, t_2 < t < t'_2$, the self-energy of the apparatus' function $\Psi_i(y_j, t)$ is smaller than that of $\Psi_o(y_j, t)$. Energy conservation of electrodynamics proves that this is precisely the additional energy deposited by the electron on device D. AB ignored this difference, stating "We have thus accomplished our objective of showing that when the source of potential is taken into account quantum-mechanically, we obtain the same result as that given in our first paper, where the potential was assumed to be a specified function of space and time" (see [83], p. 1519). The counterexample described here denies this assertion and proves that AB's electric effect violates energy conservation. Indeed, as stated above, the electric AB effect has never been detected experimentally [84]. The discussion of this section provides a theoretical explanation for this failure.

> *Conclusion: AB fail to substantiate their claim, and their electric AB effect violates the fundamental law of energy conservation [85].*

The next subsections discuss the magnetic AB effect and explain why the WKB approximation that is utilized by AB does not apply to the electric AB problem. The fundamental role of an inert source of the external field provides coherent explanations for the experimentally confirmed magnetic AB effect and for the failure of the electric AB effect.

8.8.2 The Magnetic AB Effect – Topology is not Necessary

> *Claim No-AB2:* A multiply connected region is not an essential requirement of the magnetic AB effect.

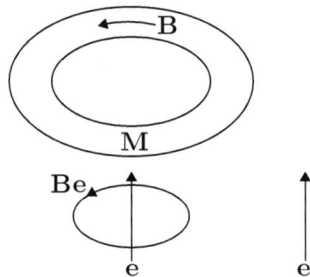

Figure 8.6: *Tonomura experiment.*

Proof: Figure 8.6 describes the magnetic AB Tonomura experiment [86]. **M** is a circular ring that is a single domain magnet, and **B** denotes its magnetic field. The magnetic ring is embedded in the (x, y) plane. An electronic beam moves in the z-direction. **e** denotes an electron of the beam that passes through the inner part of the magnetic ring **M**, and **Be** denotes the magnetic field of this electron. In addition, **e'** denotes an electron that moves along a trajectory that passes through the outer region of the magnetic single domain.

The dual nature of the two-body electromagnetic interaction is the primary point that denies the AB topological field-free region. Two completely equivalent calculations are as follows:

- One may consider the field-free region of an electron of the beam and argue that the interaction is $e\,\boldsymbol{v}\!\cdot\!\boldsymbol{A}$, where $\boldsymbol{v}$ denotes the electron's velocity and A denotes the vector potential of the magnet $\mathbf{M}$. The integral of this quantity yields the additional phase. This form apparently proves the AB claim.

- In contrast, one may argue that the interaction is the magnetic moment $\boldsymbol{\mu}$ of an atom of the single domain that interacts with the magnetic field $\mathbf{Be}$ of the moving electron $-\boldsymbol{\mu}\cdot\boldsymbol{Be}$ (see [28], p. 186). The integral of this quantity yields the additional phase. In the second picture, the field $\mathbf{Be}$ is associated with the moving electron, and *there is no field-free region.*

Obviously, a fundamental physical property should not disappear if an alternative legitimate calculation is carried out. Hence, the entire AB concept collapses. The legitimacy of the second calculation proves the claim No-AB2 – namely, the topological field-free region is not an essential element of the magnetic AB effect.

8.8.3 The Magnetic AB Effect – Topology is not Sufficient

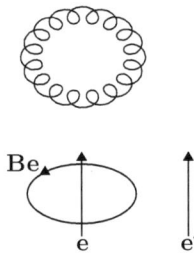

Figure 8.7: *A variation on the Tonomura experiment.*

Figure 8.7 describes a classical analog of the Tonomura experiment of the magnetic AB effect [86] (see fig. 8.6). The single domain of the permanent magnetic ring of Fig. 8.6 is replaced by a closed pipe that encompasses a circular ring. The pipe is made of an insulating material, and it contains a positively charged incompressible liquid that flows frictionlessly along the pipe. The outer part of the pipe is covered uniformly with a negative charge that neutralizes the charged liquid. Therefore, the electric field outside the pipe vanishes. The laws of electrodynamics prove that the ring's magnetic field is the same as that of the permanent magnet of the Tonomura experiment. Hence, the ring of Fig. 8.7 is a classical analog of the magnet of Fig. 8.6.

Like in the experiment described in Fig. 8.6, **e** denotes an electron of the beam that passes through the inner part of the magnetic ring, and **Be** denotes the magnetic field of this electron. In addition, **e**' denotes another trajectory of the beam that moves through the outer region of the magnetic ring. To emphasize the argument, let us change the beam's structure. Here, the beam is chopped into small sections, ensuring that no more than one electron is at the ring's vicinity at any given time.

The charge to mass ratio of the liquid that flows along the pipe is minuscule. Hence, the ring can be regarded as a classical object. Here, the time variation of the magnetic field **Be** of the electron that passes through the inner part of the ring induces an electric field that exerts a force on the moving charged liquid. Hence, the energy of the moving liquid changes.

The symbols ρ and v denote the linear charge density and the velocity of the liquid that flows along the pipe, respectively. Below, the term traveling electron is used to denote the electron for which the interference is analyzed. Quantities of traveling electrons are denoted by the subscript (e). Other quantities pertain to the coil or its relevant parts.

Let us calculate the rate of phase accumulated. The Lagrangian of this system is

$$\mathcal{L}_{total} = \mathcal{L} + \mathcal{L}_{(e)} + e\mathbf{v}_{(e)} \cdot \mathbf{A}. \tag{8.27}$$

The terms on the right side represent the Lagrangian of the coil, the Lagrangian of the traveling electron and coil-electron interaction, respectively.

The calculations take a simpler form if the coil is regarded as a dense assembly of identical circular closed loops of the pipe, each of which contains the same charged liquid that flows at the same velocity v. Let us examine the interaction between the traveling electron and one loop (see Fig. 8.8). The result for the entire coil will be derived from this analysis.

The loop's vector potential is obtained from the integration on the charged liquid that flows along the loop:

$$\mathbf{A} = \oint \frac{\rho v}{r} d\mathbf{l}, \tag{8.28}$$

where r denotes the distance from the line element $d\mathbf{l}$ to the field point where **A** is calculated, and as stated above, ρ denotes the

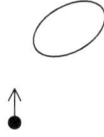

Figure 8.8: *Traveling electron and one loop.*

linear charge density. Let us use the following well known two-particle relation

$$e_1 \cdot \mathbf{v}_1 \cdot \mathbf{A}_2 = e_1 e_2 \mathbf{v}_1 \cdot \mathbf{v}_2 / r_{12} = e_2 \mathbf{v}_2 \cdot \mathbf{A}_1. \tag{8.29}$$

In this way, the interaction term of (8.27) is cast in the following form:

$$e\mathbf{v}_{(e)} \cdot \mathbf{A} = e\rho v \oint \frac{\mathbf{v}_{(e)} \cdot d\mathbf{l}}{r} \tag{8.30}$$

Now, because of the insulating material of the pipe, the charge that covers it is static throughout the experiment. Therefore, *it does not adjust its position, and it does not screen the fields of the traveling electron.* This means that the kinetic energy T and the associated Lagrangian of the rotating liquid may change during the process. For the time instant t, one uses vector analysis and Maxwell equations and finds the following change in the kinetic energy of the rotating charged liquid

$$
\begin{aligned}
\Delta T &= \int_{-\infty}^{t} \rho v [\oint \mathbf{E}_{(e)} \cdot d\mathbf{l}] dt \\
&= \int_{-\infty}^{t} \rho v [\int_{S} (\nabla \times \mathbf{E}_{(e)}) \cdot d\mathbf{s}] dt \\
&= -\int_{-\infty}^{t} \rho v [\int_{S} \frac{\partial \mathbf{B}_{(e)}}{\partial t} \cdot d\mathbf{s}] dt \\
&= -\rho v \int_{S} \mathbf{B}_{(e)} \cdot d\mathbf{s} \\
&= -\rho v \int_{S} \nabla \times \mathbf{A}_{(e)} \cdot d\mathbf{s} \\
&= -\rho v \oint \mathbf{A}_{(e)} \cdot d\mathbf{l} \\
&= -e\rho v \oint \frac{\mathbf{v}_{(e)} \cdot d\mathbf{l}}{r} \tag{8.31}
\end{aligned}
$$

This calculation proves that (8.30) and (8.31) cancel each other out. Thus, no phase shift is found for one loop of current. The same argument holds for every loop that makes the classical magnet that is discussed herein. It follows that the entire coil does not contribute to the phase.

The present experiment contains the same multiply connected field-free region as that of the Tonomura experiment. However, as stated above, an examination of (8.30) and (8.31) proves that their contributions to the rate of phase accumulated cancel each other out. Consequently, unlike the inert single domain used in the Tonomura experiment, a classical magnetic source does not alter the interference pattern.

Two conclusions can be inferred from the discussion presented above, which are as follows:

> 1. *The existence of an AB phase shift crucially depends on a source that behaves as an inert object throughout the experiment.*
>
> 2. *Contrary to the AB assertion, the multiply-connectedness topology is not a sufficient condition for having an AB effect.*

8.8.4 The Magnetic AB Effect – The Role of an Inert Source

The analysis of subsections 8.8.1 – 8.8.3 explains why the original AB articles must have an erroneous element. This subsection aims to illuminate this point. As before, the term *traveling electron* denotes the electron that produces the interference. The AB work takes the following structure:

Err.1 In their first article [82], AB ignored the source and used a single particle wave function for the traveling electron.

Err.2 Although AB recognized the erroneous element of item Err.1, their second article [83], indicates that they failed to identify the significant role of an inert object of the source of the field, as explained below.

The reason for AB's failure is quite interesting, and it deserves an appropriate analysis. In their second article [83] AB recognized the need for a wave function of the entire system and state:

"After the experiment is over (and the interaction vanishes again), the wave function will, in general, take the form of a sum of products

$$\Psi = \sum_n \phi_n(...y_i...,t)\psi_n(\boldsymbol{x},t), \qquad \text{(AB (19))}$$

where the $\psi_n(\boldsymbol{x}, t)$ represents a set of solutions of the wave equation for the electron alone, and $\phi_n(...y_i..., t)$ a corresponding set for the source variables. If such a sum of products is necessary, then it is clear that it will be impossible to factor out a one-body Schrodinger equation applying to the electronic variables alone. As we shall see, however, because the parts of the source are so heavy, only a single such product is actually needed."

Let us examine the validity of this AB's assertion. Before doing this, it is important to state that an inert source of the fields is certainly a sufficient condition for factoring out a one-body Schrodinger equation for $\psi_n(\boldsymbol{x}, t)$ of the quoted AB expression (AB (19)). The following lines explain the erroneous element of the AB's article [83]: They use the WKB approximation and claim that for all practical purposes, the function of the source is not affected by the traveling electron. This is relatively true if the state of the source is needed. *However, this argument completely ignores the accuracy that is required for the experiment, which measures quantities pertaining to a single electron.* Evidently, a quantity – such as energy – that can be measured by a single electron is unmeasurable if it is distributed between the astronomical number of elements that compose the source of the fields.

The notion of entanglement clarifies the problem. This notion is defined as follows [87]:
"Quantum entanglement is a physical phenomenon that occurs when a pair or group of particles is generated, interact, or share spatial proximity in a way such that the quantum state of each particle of the pair or group cannot be described independently of the state of the others, including when the particles are separated by a large distance. The topic of quantum entanglement is at the heart of the disparity between classical and quantum physics: entanglement is a primary feature of quantum mechanics lacking in classical mechanics.

Measurements of physical properties such as position, momentum, spin, and polarization performed on entangled particles can, in some cases, be found to be perfectly correlated."

This definition of entanglement applies to the AB experiments. Indeed, there is an entangled state of the traveling electron and the source. Here, each term of the quoted AB sum – (AB (19)) – of the entangled system has its own phase. The accuracy of the calculations requires a correct expression for the electron's energy, as well as a unique phase of each term of the system. The law of energy conservation of electrodynamics proves that the source's energy state contributes to the phase. This is demonstrated by the analysis of the experiments of the electric AB effect described in subsection 8.8.1 and the magnetic effect in subsection 8.8.2. These counterexamples prove that the WKB approximation that AB use for the source does not correctly describe the required precision of the quantum state of the system.

A brief overview of the above issue can be given as follows: Ignoring small quantities is a well-known practice in scientific work. For example, many physicists and most engineers use Newtonian mechanics for cases where the velocity of all elements of the system is much smaller than the speed of light. AB used the Schroedinger QM and not RQM or QFT. The main point of these cases is that the quantities that are ignored are extremely small, and the final result remains practically the same if one ignores them.

AB treated the classical source as an isolated object and argued that the WKB approximation is permissible. However, this is not the case with the AB experiment, which depends on an accurate calculation of the phase of *a single electron*. Hence, the change of the classical Lagrangian of the source, $T - V$, should be accurately calculated. *Here, quantities of the order of magnitude of the energy and the phase of a single electron should not be ignored.* The counter-examples given above prove that the AB application of the WKB approximation is unjustified.

Conclusions:

AB.1 The AB application of the WKB approximation for a description of the source's state is unsuitable for the phase calculation of the entangled structure of the traveling electron and the fields' source.

AB.2 AB's work missed that an inert source (e.g. the magnet of the Tonomura experiment) is a necessary and sufficient condition for the AB interference.

AB.3 As proved in subsection 8.8.1, these erroneous AB elements violate the law of energy conservation.

8.9 Summary – the 4-Potential A_μ

At present, many textbooks regard the electromagnetic 4-potential as an important physical element. For example, the 4-potential is used in the *gauge covariant derivative*

$$D_\mu = \partial_\mu + ieA_\mu(x), \tag{8.32}$$

which can be found in many textbooks (see e.g. [14], p. 78). This expression explicitly depends on the 4-potential. For this reason, the new results about the 4-potential that are discussed in this book are summarized as follows:

A.1 Maxwell equations depend on a conserved 4-current j^μ. In VE, the Lagrangian density is a Lorentz scalar with a dimension of $[L^{-4}]$. Hence, the electromagnetic interaction term requires a contraction of j^μ with a 4-vector whose dimension is $[L^{-1}]$. This 4-vector should be written in the form $A_\mu(x)$, where x denotes the four space-time coordinates used for the quantum function $\psi(x)$. However, a requirement for such a 4-vector does not ensure that $A_\mu(x)$ is a genuine 4-vector! This book proves the relevant points outlined below.

A.2 Radiation fields and bound fields are different physical entities (see section 8.2). Hence, an expression for the 4-potential that yields these intrinsically different fields cannot be theoretically coherent.

A.3 A relativistically consistent transformation can be carried out for the 4-potential of radiation fields (see subsection 8.3.4). However, this is not a Lorentz transformation. Hence, the 4-potential is *not* a 4-vector, which means that it is an auxiliary quantity.

A.4 Several factors explain why the Lienard-Wiechert 4-potential is unsuitable for the interaction term of a charged particle Lagrangian density. These are as follows:

LW.1 The Lienard-Wiechert 4-potential is a 4-vector

$$A_\mu = e\frac{v_\mu}{R^\alpha v_\alpha}. \tag{8.33}$$

Hence, it cannot represent the A_μ of radiation fields because the latter is not a 4-vector.

LW.2 The Lienard-Wiechert 4-potential (8.33) depends on local coordinates of the Lagrangian density *and* on retarded coordinates of the radiating system. Hence, it depends on degrees of freedom that do not belong to the Lagrangian density of the interacting electron.

LW.3 A single photon is emitted from the *entire* radiating system (see subsection 8.3.3). Hence, Lienard-Wiechert single-charge expression of the 4-potential (8.33) cannot describe radiation.

A.5 The AB arguments are used as proof of the physical significance of the 4-potential (see e.g. [88], section 6). However, it was proved above (see section 8.8) that these arguments are inherently wrong. Hence, AB's work cannot substantiate the physical meaning of the 4-potential.

A.6 The gauge covariant derivative (8.32) is a dubious expression because it comprises two objects that undergo a different relativistic transformation.

> Conclusion: MLE does not use the 4-potential. Hence, in MLE, the 4-potential is an auxiliary mathematical expression. The 4-potential is explicitly used in the interaction term of VE. However, it is not a 4-vector. Therefore, it is an auxiliary expression, and the relativistic transformation of the 4-potential of radiation fields differs from the ordinary Lorentz transformation of a 4-vector.

Chapter 9

Particles and Interactions

9.1 An Introduction to Elementary Particles and Their Interactions

This section briefly describes two kinds of physical categories – massive particles and their interactions. Experimentally confirmed particles are organized in sets, where all members of each set participate in the same kinds of interactions.

Physics recognizes four kinds of forces/interactions – gravitational, weak, electromagnetic, and strong. This book does not discuss details of gravitational interaction. Analogously, quantum theories in general and the SM in particular, take the same approach: They analyze the weak, electromagnetic, and strong interactions. Below, the term *three relevant interactions* denotes these interactions. This book and the SM present different theories that apply to the same physical topics – namely, a theoretical description of the three relevant interactions. This chapter is dedicated to the general properties of the theories that are supported by this book, and alternatively, by the SM. Let us begin with an examination of an important issue: Theoretical physics uses experimental data as a clue for taking the right direction. Hence, let us see how this book and the SM use fundamental properties of the data that belong to the three relevant interactions.

One aspect of physical interactions is their theoretical proximity. Maxwell equations provide an example of the unification of electricity and magnetism, where the same set of equations describe these phenomena. It can be stated that the unification of theories of different interactions is discussed in the present literature (see, e.g., the notion "theory of everything"). The discussion that is carried out here is limited to one step in this direction. It examines how this book and the SM use fundamental data and construct a common theoretical basis for two interactions that are included in the three relevant interactions. Before doing this, let us examine the kinds of elementary spin-1/2 massive particles and their interactions.

Relying on experimental data, the elementary spin-1/2 massive particles are organized into three sets – neutrinos, charged leptons, and quarks. Neutrinos participate only in weak interactions. Charged leptons interact electromagnetically and weakly. Quarks participate in strong, electromagnetic, and weak interactions.

This book adopts the variational principle as a basis for a quantum theory of particles and their interactions. Hence, the Lagrangian density of a neutrino takes the form

$$\mathcal{L}_N = \mathcal{L}_{Self} + \mathcal{L}_{Weak}, \tag{9.1}$$

where the Lagrangian density of a particle depends on the corresponding quantum function ψ and its derivative $\mathcal{L}(\psi, \psi_{,\mu})$. The Lagrangian density that represents an interaction of a given particle takes the form $\mathcal{L}(\psi, \psi_{,\mu}, \phi, \phi_{,\mu})$, where ϕ denotes the quantum field that carries the interaction.

The Lagrangian density of a charged lepton, like that of the electron, is an extension of (9.1). Here, a term that describes the electromagnetic interaction (EM) is added as follows:

$$\mathcal{L}_e = \mathcal{L}_{Self} + \mathcal{L}_{EM} + \mathcal{L}_{Weak}. \tag{9.2}$$

By the same token, the Lagrangian density of a quark takes the form

$$\mathcal{L}_Q = \mathcal{L}_{Self} + \mathcal{L}_{Strong} + \mathcal{L}_{EM} + \mathcal{L}_{Weak}. \tag{9.3}$$

Details of this approach are discussed in the next chapters.

9.2   Unification of a Pair of Interactions

Like in other cases of theoretical work in physics, an adequate examination of the relevant data is a good starting point. Tables 9.1 and 9.2 demonstrate an inherent difference between this book and the SM. In each table, each row shows one of the three relevant interactions. The three rightmost columns, each shows a fundamental physical effect. Two rows pertaining to a pair of interactions that have a common theoretical basis are grouped together. The entries "YES/NO" denote the case where every process dominated by a specific interaction does or does not have the experimental property of the corresponding effect.

Table 9.1: Interaction Unification in the RCMT. Conserv. denotes Conservation.

Theory	Interaction	Parity Conserv.	Flavor Conserv.	Photon Interaction
RCMT	Strong	YES	YES	YES
	EM	YES	YES	YES
DDWIT	Weak	NO	NO	NO

Table 9.2: Interaction Unification in the SM.

Theory	Interaction	Parity Conserv.	Flavor Conserv.	Photon Inter.
QCD	Strong	YES	YES	YES
Electroweak	EM	YES	YES	YES
	Weak	NO	NO	NO

Table 9.1 shows that the theories supported by this book adhere to the meaning of fundamental experimental data. The strong and electromagnetic interactions are grouped together, and the data firmly supports this issue. Table 9.2 proves that the case of the SM is completely different. Its electroweak theory unifies the electromagnetic and weak interactions. Hence, *the electroweak theory must use a mathematical framework that describes intrinsically different physical data.* Furthermore, the SM ignores the significant similarity between strong and electromagnetic interactions. For example, ample experimental data are supporting a hard photon-hadron interaction. SM textbooks simply do not discuss this kind of data.

This book devotes special attention to the form of the theories that describe how a Dirac particle is affected by each of the interactions shown in (9.3). This expression represents a Lagrangian density where – like in the electromagnetic case – each of the three relevant interactions is represented by one specific term.

The data that are described in Tables 9.1 and 9.2 can certainly convince an unbiased reader that this book supports a solid theoretical approach, whereas the SM is an assembly of groundless theories. The rest of this book shows *many* examples that substantiate this opinion.

Chapter 10

Magnetic Monopoles and Strong Interactions

10.1 General Strong Interaction Properties

This chapter discusses processes that are determined by strong interactions and compares the relevance of two theories that aim to describe these interactions – QCD and the RCMT. Electromagnetic processes, where the coupling constant $e^2 \simeq 1/137$, are relatively weak. This feature facilitates the calculation of processes where electrodynamics plays a dominant role. This is not the case with strong interactions.

Table 10.1 shows corresponding quantum states that are determined by electromagnetic and strong interactions and their relative intensities. The penultimate column denotes the total energy of the state, and the last column denotes the ratio between the relevant excitation energy and M_0. The rows of this table show energy states of the positronium, which are determined by electromagnetic interactions and corresponding $\bar{q}q$ states of $I = 1$ mesons where the valence quarks are of the u or d flavor. The latter states are determined by strong interactions. Numerical values are written in an appropriate approximation.

The table shows that the ratio between the electromagnetic interaction energy and the total mass of the positronium's ground state is extremely small. Therefore, a positronium's state can be regarded as a minor perturbation of a two free-particle state. Here, relativistic effects are quite small, and numerical perturbation cal-

117

Table 10.1: Properties of interacting particles. Notation: Int. –
Interaction; EM – electromagnetic; St. – Strong; n, ℓ, J, π –
Spectroscopic symbols; M_i – The total mass of the ith state; M_0 –
The total mass of the ground state; Posit. means positronium.

Name	Int.	n	ℓ	J^π	M	$(M_i - M_0)/M_0$
Posit.	EM	1	0	0^-	$(2m_e - 6.8)$eV	0
Posit.	EM	1	0	1^-	$(2m_e - 6.8)$eV	$\simeq 0$
Posit.	EM	2	1	0^+	$(2m_e - 1.7)$eV	1/200000
Posit.	EM	2	0	0^-	$(2m_e - 1.7)$eV	1/200000
π	St.	1	0	0^-	140 MeV	0
ρ	St.	1	0	1^-	775 MeV	4.5
A_0	St.	2	1	0^+	980 MeV	6
A_1	St.	2	1	1^+	1230 MeV	7.8
$\pi(1300)$	St.	2	0	0^-	1300 MeV	8.3
A_2	St.	2	1	2^+	1317 MeV	8.4

culations are successful. In contrast, strong interactions cancel out
most of the self-mass of the components, meaning that the relativistic effects are huge. Hence, a numerical solution of a quantum state
that is determined by strong interaction is a difficult assignment.

Another kind of experimental data supports the above conclusion. The proton's state is characterized by the three uud valence
quarks. In addition to these quarks, the proton state comprises a
non-negligible number of quark-antiquark pairs of the u, d, s flavor
(see subsections 10.5.15, 10.5.16). Hence, the mathematical equation of the proton is much more complicated than a three-body
problem would be. The additional quark-antiquark pairs in the
proton have been found in experiments. Moreover, the state of
every hadron is determined by strong interactions. Therefore, the
expectation of the existence of additional quark-antiquark pairs in
the state of every hadron is justifiable. For these reasons, the mathematical problem of hadronic states is a complicated many-body
problem.

The objective difficulty of strong interaction calculation entails
that the comparison between QCD and the RCMT discussed in this
book uses qualitative data. Nevertheless, qualitative arguments
can be effective for this aim. For example, let us compare the state
of the proton's antiquarks with that of its quarks. In principle, one
may expect that one of the following possibilities holds: Antiquarks
are enclosed in a clearly larger volume than that of quarks; antiquarks are enclosed in about the same volume as that of quarks;
antiquarks are enclosed in a clearly smaller volume than that of

quarks. The data show that the proton's antiquarks are pushed to its peripheral regions (see subsection 10.5.14). Hence, a strong interaction theory should provide qualitative arguments that explain this antiquark property. Analogous problems are discussed in this chapter.

10.2 QCD has been Constructed on an Erroneous Basis

Consider the $\Delta^{++}(1232)$ baryon. SM textbooks claim that each of the single-particle wave functions of its three uuu valence quarks is a ground state s-wave based on the following argument: The $\Delta^{++}(1232)$ is the lightest state of the uuu quarks. Therefore, each of its three uuu quarks should be in a single particle ground state s-wave – namely, a symmetric spatial state. Furthermore, the quantum numbers of the $\Delta(1232)^{++}$ are $J^{\pi} = 3/2^{+}$, which is a symmetric spin state. Hence, the data demonstrate a fiasco of the Fermi-Dirac statistics of ordinary QM (see e.g. [38], p. 5).

However, this is an incomplete argument because it resembles the case where one examines just a single tree and ignores the forest. Indeed, the next lines describe the complete relevant data. The four $\Delta(1232)$ baryons

$$(\Delta^{-}, \Delta^{0}, \Delta^{+}, \Delta^{++}), \qquad (10.1)$$

are members of the lightest isospin quartet of the Δ baryons. Hence, the analysis should go as follows: Figure 10.1 shows six baryons that are organized in two isospin multiplets. The two nucleons are in $I = 1/2$, and the four $\Delta(1232)$ baryons are in $I = 3/2$. Each horizontal segment represents a baryon. The baryon's name is written below the segment, and its valence quarks are written above the segment. The goodness of the isospin notion indicates that all members of the $\Delta(1232)$ baryons have the same space, spin, and isospin symmetry. For example, Wong uses the nuclear physics terminology[1] and states: "Physically, it means that the wave function of a state with a definite isospin T is unchanged if we replace some of the protons by neutrons, and vice versa" (see [8], p. 73). Furthermore, Fig. 10.1 clearly shows that the Δ^{0} and the Δ^{+} baryons are *excited states* of the neutron and the proton, respectively. As

---

[1] Readers should note that the isospin group of nuclear physics is identical to that of particle physics. The sole difference between them is the notation: Nuclear physicists use the letters t, T, whereas particle physicists use the letters i, I.

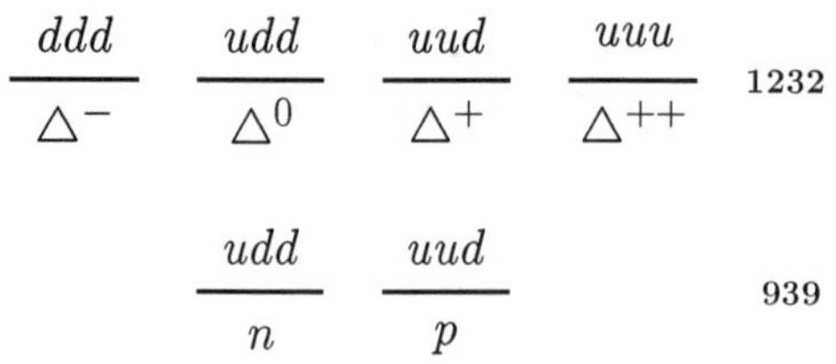

ddd	udd	uud	uuu	
Δ^-	Δ^0	Δ^+	Δ^{++}	1232

udd	uud	
n	p	939

Figure 10.1: *Two isospin multiplets and their energy (in MeV).*

excited states, the laws of ordinary QM prove that the Δ^0 and the Δ^+ baryons should have excited space-spin states. This conclusion holds for every member of the isospin quartet (10.1). Hence, the space-spin part of the single-particle wave function of the Δ^{++} uuu quarks is *not* a ground state s-wave, and its entire state is consistent with the laws of ordinary QM. Points A and B below refer to the Δ^{++} baryon, which is the lightest baryon of this kind, and show where QCD has gone wrong.

> A. QCD aims to provide an answer to the wrong question: What are the quark's additional degrees of freedom that enable the three uuu valence quarks of the Δ^{++} baryon be in the $1s^3$ ground state without violating the Pauli exclusion principle?

> B. This section proves that the right question is as follows: What is the Δ^{++} structure, and what are the laws of strong interactions, that put the 3 uuu valence quarks in a state that is not a $1s^3$ ground state?

An analogous argument holds for the **8** and **10** representations of the SU(3) group, which includes the corresponding baryons of Fig. 10.1. For example, the $J^\pi = 3/2^+$, $\Sigma^+(1385)$ baryon of **10** is an excited state of the $J^\pi = 1/2^+$, $\Sigma^+(1189)$ baryon of **8**, and so on. For this reason, the ground state $J^\pi = 3/2^+$ of the sss quarks of the Ω^- baryon is *not* a single particle s-wave state. Hence, the laws of ordinary QM also explain the Ω^- state. Contrary to this conclusion, H. Fritzsch – one of the QCD pioneers [89] – claims that the three quarks of the Ω^- baryon should be in a single particle s-wave state and that this assumption was the specific reason for the QCD construction [90].

> *Conclusion: There is no need for QCD, and the basis for its construction is erroneous.*

It is interesting to note that the effect called *the proton spin crisis* emanates from another erroneous QCD idea of point A of this subsection: A single configuration cannot describe a baryonic state. This issue provides another proof for the claim that QCD has been constructed on an erroneous basis (See also section 7.5, on p. 78, and subsection 10.5.18 on p. 152).

10.2.1   Spin-Dependence of Interactions

Section 7.5, p. 78, explain why a multi-configuration structure is required for a baryon description. For the simplicity of the discussion, this subsection mentions only one configuration of each state.

Let us examine another aspect of the Δ baryons and the spin dependence of electromagnetic and strong interactions. In the case of electromagnetic interactions of an atomic state, we have the Hund rule that says that the lowest energy state of a given electronic configuration is its highest spin state (see [27], p. 226; [73], p. 702). Specifically, if the spins are parallel, the binding energy is stronger.

In the case of the positronium, which is an electron-positron bound state, the interaction changes sign in relation to the electron-electron interaction of atomic states. Here, the ground state is the spin $= 0$ state, and the spin $= 1$ state takes a somewhat higher energy level [91]. Evidently, the change of the interaction sign reverses the Hund's order of energy states.

Another aspect of the spin's role in electromagnetic interactions is that its relative size increases with the increase of energy (see [23], pp. 192, 193). Hence, if RCMT is right, then the spin effects should play a significant part in strong interactions.

Strong interactions are much stronger than electromagnetic interactions are, and their spin-dependent effects are more dramatic. Let us examine the two mesons – that is, π, $j^{\pi} = 0^-$, $m = 140$ MeV and ρ, $j^{\pi} = 1^-$, $m = 775$ MeV. As in the case of the positronium, the π mesons with the lower spin state are bound more strongly compared with the ρ mesons with the higher spin state. As expected for mesons, the energy difference between these two spin arrangements is large.

The isospin quartets of the Δ baryons comprise the Δ^{++}, which has three uuu quarks. This means that if RCMT is right, the system is analogous to atomic electrons. Its states demonstrate the quarks' version of the Hund rule (see table 10.2), but the corresponding energy differences are much greater than those of the elec-

Table 10.2: Data of Δ Baryons

Name	Quarks	J^π	Mass (MeV)
$\Delta(1232)$	uuu	$\frac{3}{2}^+$	1232
$\Delta(1600)$	uuu	$\frac{3}{2}^+$	1570
$\Delta(1620)$	uuu	$\frac{1}{2}^-$	1610
$\Delta(1700)$	uuu	$\frac{3}{2}^-$	1710

tromagnetic case. Here, the lowest and the second-lowest states are $\frac{3}{2}^+$, meaning that the three Dirac particles have the highest spin.

The difference between the $\Delta(1232)$ and the $\Delta(1600)$ is a radial excitation. Hence, the gain of the binding energy of the higher spin is stronger than the additional energy of the radial excitation!

It should be emphasized that the single configuration terminology used here is just for the simplification of the arguments, as stated at the beginning of this subsection. Indeed, a single configuration does not correctly describe the electronic and quark states (see section 7.5 on p. 78 and subsection 10.5.18 on p. 152).

10.3 The Regular Charge-Monopole Theory

Quarks are spin-1/2 particles that carry an electric charge. As such, one may expect that the quark's electromagnetic interaction should be described by the Dirac Lagrangian density. To assess this claim, let us again write the Dirac Lagrangian density (3.33) in the form where the electromagnetic interaction term stands separately:

$$\mathcal{L}_D = \bar{\psi}[\gamma^\mu i\partial_\mu - m - e\gamma^\mu A_\mu]\psi. \tag{10.2}$$

The electromagnetic interaction term of (10.2) is called "minimal interaction." One may examine this minimal interaction and the successful properties of QED and define the strong interaction problem as given below.

> **The Strong Interaction Problem:** What term (or terms) should be added to the Lagrangian density (10.2) to correctly describe strong interactions?

The solution to this problem is found in a surprising place.

10.3.1 An Introduction to the Monopole Problem

Several elements of the RCMT are discussed in subsection 3.6.2, p. 30. This theory is used for a description of many strong interaction effects.

An important difference between an electric charge and a monopole is that the electric charge has been found in laboratories, whereas a monopole has not yet been found. Therefore, the monopole issue is a subject of theoretical work. Here, one relies on the basic properties of electrodynamics, and in particular, on the well-known relationships between electric charge and electromagnetic fields. Thus, one expects that analogous relationships exist between monopoles and electromagnetic fields. Below, this correspondence is called charge-monopole duality and the duality notion can be defined mathematically (see subsection 3.6.2 on p. 30).

The lack of experimental monopole data means that a brief survey of the literature is a good starting point. Let us take two well-known textbooks on classical electrodynamics – those of Landau and Lifshitz [3] and Jackson [28]. One finds that Landau and Lifshitz do not mention monopoles at all, whereas Jackson dedicates only a tiny portion of his book to monopoles.

Wikipedia is a well-known source of information. According to its principles, the site describes ideas that are consistent with the current consensus. The Wikipedia's electric charge entry describes properties of one kind of physical entity (which can be either positive or negative). In contrast, the Wikipedia's monopole entry describes several kinds of monopoles.

The previous lines show that the present status of the electric charge is completely different from that of the monopole. This difference is incompatible with the charge-monopole duality, which states that a monopole theory should be analogous to ordinary electrodynamics. In particular, if nature shows one kind of charge, then there should only be one kind of monopole. Therefore, it is clear that the present consensus about monopoles has not given the last word on this subject. For this reason, monopoles look like a promising subject for theoretical research.

10.3.2 A Definition of Monopoles

The theoretical way towards the definition of electric charge is based on experimental data. An electric charge can be regarded as a source of electromagnetic fields. In contrast, it can also be defined as the entity by means of which electromagnetic fields exert force on a massive charged particle. These issues are just two facets of the same thing, and Maxwellian electrodynamics unifies them coherently.

The lack of experimental monopole data means that the monopole definition cannot follow the charge's path. Hence, it should depend on theoretical arguments. Let us rewrite the duality relations of subsection 3.6.2:

$$\boldsymbol{E} \to \boldsymbol{B}, \quad \boldsymbol{B} \to -\boldsymbol{E}, \tag{3.28}$$

$$e \to g, \quad g \to -e \tag{3.29}$$

(see [28], pp. 251-252, [32], p. 1363). The transformations of the electromagnetic fields (3.28) can be put in a tensorial form as

$$F^{\mu\nu} \to F^{*\mu\nu}, \quad F^{*\mu\nu} \to -F^{\mu\nu}, \tag{3.30}$$

where $F^{*\mu\nu} = \epsilon^{\mu\nu\alpha\beta} F_{\alpha\beta}$. The transformations (3.28) and (3.29) are sometimes called duality rotations by $\pi/2$ (see [28], p. 252).

Subsection 3.6.2 explains why the RCMT is a reasonable approach for a physically meaningful monopole theory. The applicability of the RCMT to strong interactions is discussed in this chapter.

10.3.3   A Regular Charge-Monopole Theory

Relying on the correspondence principle, one derives self-evident requirements that are imposed on the RCMT:

CMT.1 For systems of charges without monopoles, the RCMT must agree with the ordinary Maxwellian electrodynamics.

CMT.2 For systems of monopoles without charges, it must agree with the dual theory described above.

Hence, one must solve the problem below.

> Problem: What is the form of the Lagrangian density that yields a RCMT that abides by the CMT.1, CMT.2 requirements?

This problem has been solved, and [36] describes the solution in detail. The main results of the solution are as follows:

RCMT.A Charges do not interact with bound fields of monopoles.

RCMT.B Monopoles do not interact with bound fields of charges.

RCMT.C Radiation fields of the system are identical and charges, as well as monopoles, interact with them.

RCMT.D Unlike in the Dirac monopole theory, the strength of the elementary monopole unit is a free parameter.

RCMT.E Unlike in the Dirac monopole theory, the RCMT is free of irregular strings and of the artificial limitation that forbids an electric charge from being on the string's space-time points (see [28], pp. 251-260).

It is interesting to point out that the RCMT's distinction between bound fields and radiation fields of items RCMT.A-RCMT.C is compatible with the proofs of section 8.2, p. 84. These proofs show that these fields are inherently different physical objects. For a given system, let $A_{(ew)\mu}$ denote the 4-potential that yields the fields of charges and radiation fields and $A_{(mw)\mu}$ denote the 4-potential that yields the fields of monopoles and the radiation fields of this system. Here, the symbols e and g denote the charge and the monopole strength of a Dirac particle, respectively, and the symbol w denotes radiation fields. (In subsection 8.3.3, it is proved that the emission of radiation fields cannot be assigned to a specific charge. This outcome is consistent with this notation, where radiation fields are treated as independent physical objects.) The form of the Dirac Lagrangian density of a particle that carries charge e and monopole strength g is

$$\mathcal{L}_{CM} = \bar{\psi}[\gamma^\mu i\partial_\mu - m - e\gamma^\mu A_{(ew)\mu} - g\gamma^\mu A_{(mw)\mu}]\psi. \quad (10.3)$$

Therefore, requirements CMT.1 and CMT.2 are satisfied. Thus, if monopoles do not exist, $g = 0$, and the radiation fields are those emitted by systems of charges. Therefore, (10.3) reduces to (10.2). Analogously, one finds that for a system without charges, (10.3) reduces to the dual theory of monopoles.

Properties RCMT.A – RCMT.C have an important realization in the experimental data. Thus, electrons – namely, pure electric charges – do not participate in strong interactions (see [38], p. 2). In contrast, real photons do interact strongly with the nucleon constituents – namely, quarks [39]. This is precisely a description of the data of strong and electromagnetic interactions. Electrons and other charged leptons do not carry monopoles, whereas quarks carry both monopoles and charges. (A particle of this kind is called a dyon.)

10.3.4   Important RCMT properties

The important RCMT properties and its applicability to hadronic states and their interactions are summarized below. They are helpful for a reference in later analysis.

RCMT.1 The charge-monopole duality relations mean that strong interactions are analogous to electromagnetic interactions. The experimental evidence of Table 9.1 on p. 115 supports this assumption.

RCMT.2 Items MR.1-MR.3 in subsection 3.6.2 and the similarity between the strong and electromagnetic interactions of RCMT.1 yield these claims: Purely charged particles (like charged leptons) do not participate in strong interactions. For an experimental confirmation of this point, see [38], (p. 2). In contrast, real photons do participate in strong interactions. For an experimental confirmation of this point, see [39].

RCMT.3 The structure of a baryon is similar to that of an atom: Like electrons, quarks are ordinary Dirac particles, and a baryon has a core comprising an object that carries a positive monopole strength and some closed shells of u and d quarks. The overall monopole strength of this core is $3g$. A baryon also has three valence quarks and an experimentally determined number of $\bar{q}q$ pairs, primarily of the u, d and s flavors. Each quark carries a monopole strength $-g$. Hence, a baryon is neutral respecting the monopole strength.

Conclusion: This baryonic structure and the similarity between the strong and electromagnetic interactions of item RCMT.1 mean that the nuclear force is analogous to the molecular force. Moreover, the mesonic structure is analogous to that of the positronium.

RCMT.4 It is well known that atomic states are determined by the electrodynamic laws of QM. The first Hund law says that for the same configuration, a maximal spin yields a lower energy state ([27] p. 226, and notice the ‡ comment).

High-energy electron scattering provides another aspect of high-energy spin-dependent magnetic interaction. Here the Rosenbluth formula shows that spin-dependent processes become dominant with the increase of energy (see e.g. [79], p. 172).

The RCMT says that strong interactions are analogous to electromagnetic interactions. An examination of the low energy mesons supports this issue, and the spin-dependent interaction is clear. Thus, pions have $j^\pi = 0^-$, and their mass is less than 140 MeV; ρ mesons have $j^\pi = 1^-$, and their mass is around 770 MeV [29]. This means that for $\bar{q}q$ systems of mesons, lower energy goes with anti-parallel spin states. By the same token, lower energy states of a qq interaction are related to parallel spin states.

Item RCMT.4 on p. 126 is relevant to the $j^\pi = 3/2^+$ of the Ω^- baryon. It shows that the energy of the π, ρ mesons (a $\bar{q}q$ system) strongly supports the lower energy of qq parallel spin states.

10.4   The Main Elements of QCD

QCD is the SM sector that aims to describe the structure of hadrons and the dynamic laws that determine their state. The existence of quarks was established in the 1960s, and QCD regards them as the building blocks of hadrons. However, as explained in section 10.2 on p. 119, QCD has been constructed on an erroneous basis. Consequently, this theory probably contains other erroneous elements.

QCD assumes that the hadronic structure is as follows:

1. Mesons are bound states of $\bar{q}q$.

2. Baryons are bound states of qqq, which are called valence quarks. In addition to these quarks, QCD allows baryons to comprise a probability of $\bar{q}q$ pairs, where members of each pair have the same flavor. In addition to quarks, baryons comprise gluons.

The QCD laws take a mathematical extension form of the electromagnetic force. According to QED, the commutative U(1) group pertains to electrodynamics. Applying an extension of this concept, one can use the non-commutative SU(3) group and build a theory comprising three kinds of objects that are analogous to the electric charge. These objects are called colors, and they are named Red, Green, and Blue (RGB). The color degree of freedom is an element of the QCD structure. (The QCD colors are just names that have nothing to do with the ordinary colors seen by human eyes.)

The color degree of freedom "corrects" the QCD problematic points of 10.2 on p. 119, and it "restores" the Fermi-Dirac statistics of quark states.

The non-commutative SU(3) group of QCD means that its mathematical structure is not simple. For further reference, let us note some of its features:

A. The QCD color degree of freedom adds three additional quantum states that enable the three $1s^3$ valence quarks of the Δ^{++} to take an antisymmetric state and abide by the Pauli exclusion principle.

B. QCD introduces a specific hypothesis: A system of quarks cannot be detected as a specific free particle unless its overall color is white, namely it comprises an equal amount of $R, G,$ and B.

C. Asymptotic freedom is a peculiar QCD feature. It asserts that the coefficient of the QCD interaction intensity *decreases* with the decrease of the quark-quark distance. This feature means that the intensity of this coefficient *increases* with the increase of the quark-quark distance. This counterintuitive feature is discussed in several places in this book.

D. A high-energy elastic proton-proton scattering event is a non-negligible effect (see sections 10.5.11, 10.5.12 that begin on p. 142). The dynamic attributes of this scattering are analogous to the electron scattering in Fig. 8.1 on p. 86. In both cases, the 4-momentum that is transferred in the process is space-like, meaning that a virtual particle is involved. In contrast, QCD assigns about one-half of the proton mass to gluons. Hence, the gluons that are enclosed inside the proton are supposed to be genuine particles where the 4-momentum is time-like.

It is not clear how can QCD account for the existence of real gluons and virtual gluons. However, it is shown later in this chapter that QCD faces many kinds of explicit contradictions.

10.5   The RCMT Versus QCD

Up to this point, one may cast doubt on QCD because of general principles in terms of its construction on erroneous arguments (see section 10.2, on p. 119). This point does not decisively deny

QCD because one may argue that a correct theory can *luckily* be constructed on such a basis.

The RCMT form of the Lagrangian density of strong and electromagnetic interactions (see (10.3) on p. 125) is another general argument because the RCMT introduces an interaction term that is analogous to the minimal interaction of electrodynamics. Hence, Occam's razor arguments support RCMT and cast doubt on QCD.

It is interesting to note a somewhat amusing aspect of QCD: M. Gell-Mann was a key figure in the QCD construction. According to Wolfram, Gell-Mann told him "after hearing a talk I gave on QCD that I should work on more worthwhile topics" [92]. I am quite sure that Einstein never said something like this to anyone working on SR or general relativity. The same is true with Dirac and his quantum equation. In contrast, as stated above, the process called renormalization is an element of QED, and Feynman, a key QED figure, described renormalization as "a dippy process" (see [77], p. 128). The rest of this chapter describes many kinds of experimental data that support the RCMT and deny QCD.

10.5.1 The Hard Photon-Nucleon Interaction

Experiments that have been carried out many years ago prove that "the limiting photon total cross-sections on neutrons and protons are nearly the same, indicating that the photon interaction does not depend primarily on the charge of the target" (see [39], p. 269). The SM has no explanation for these data, and SM textbooks do not discuss this issue.

Electrons, photons, and nucleons are well-known particles and experimental physicists have provided data of two-body scattering of each pair of these particles. Figure 10.2 describes the unfortunate SM status of this issue. The figure shows scattering experiments

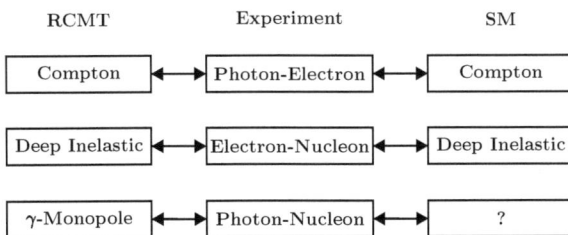

RCMT	Experiment	SM
Compton	Photon-Electron	Compton
Deep Inelastic	Electron-Nucleon	Deep Inelastic
γ-Monopole	Photon-Nucleon	?

Figure 10.2: *Three particles: scattering data of each pair and their interpretation by RCMT and SM.*

that have been carried out for every pair of these particles. The photon-electron scattering (called Compton scattering) is discussed in relevant textbooks on QED (see e.g. [14], pp. 158-167; [38], pp. 141-144). The same is true with the electron-nucleon deep inelastic scattering (see e.g. [14], pp. 475-480, 555-563 and 621-647; [38], chapter 8). The two kinds of electron scattering are electromagnetic processes that have adequate explanations. In contrast, SM textbooks neglect the case of a hard photon scattered on a nucleon. Here, experimental data do exist, but SM textbooks ignore the effect.

It turns out that subsection RCMT.2 of section 10.3.4 mentions important RCMT properties which provide a straightforward interpretation for this effect: Quarks are monopoles where the unit monopole strength is much greater than that of the electron's electric charge. Hence, the RCMT's property 10.3.4 says that the intensity of hard photon interaction with a proton should be about the same as its interaction with a neutron.

> Conclusion: The RCMT provides an immediate explanation for the hard photon-nucleon interaction. SM has no explanation for this effect, and textbooks in the field ignore this topic. Furthermore, this SM problem is not mentioned on the Wikipedia list of unsolved problems in particle physics [93].

10.5.2 The Strange Episode of the Vector Meson Dominance Idea

The data mentioned in the previous section cannot be explained by the conventional properties of electrodynamics. Historically, the photon has been regarded as a pure electromagnetic particle that is associated with Maxwellian radiation fields. Therefore, the new data about the hard photon-nucleon interaction is inconsistent with the prevailing theory of that time. An idea called vector meson dominance (VMD) has been suggested to explain the new data. Similar ideas are called modified VMD (MVMD), vector dominance model (VDM), and hadronic structure of the photon. The primary element of these ideas states that the wave function of an energetic photon takes the form (see [39], p. 271)

$$| \gamma > = c_0 | \gamma_0 > + c_h | h >, \qquad (10.4)$$

where $| \gamma >$ denotes the wave function of a physical photon, $| \gamma_0 >$ denotes the pure electromagnetic component of a physical photon, and $\mid h >$ denotes its hypothetical hadronic component. Moreover, c_0 and c_h are appropriate numerical coefficients; c_h is insignificant in the case of a soft photon. On the other hand, a hard photon is described by a non-negligible value of c_h. It means that "The hadronic contribution for low-energy photons is insignificant. The high-energy photon is accompanied by a hadron cloud that leads to observable effects" (see [94], p. 319). Evidently, the second term on the right-hand side of (10.4) is an example of an ad hoc error correction. Indeed, it has been suggested *after* the new photon-nucleon data have been measured, and it aims to explain them by changing an established theoretical expression. The following arguments prove that the VMD relation (10.4) is inconsistent with fundamental physical principles.

Relation (10.4) is inconsistent with Wigner's analysis of the Unitary Representations of the Inhomogeneous Lorentz Group (see section 3.7 on p. 37). This analysis proves that a massive quantum particle has a well-defined mass and spin whereas a massless particle has two components of helicity. It means that $\mid \gamma_0 >$ of (10.4) has only two values of the z-component of its spin, whereas $\mid h >$ of (10.4) has three values of the z-component of its spin. This is a contradiction because a free quantum particle has a well-defined spin J and spin projection J_z (see section 3.7 on p. 37).

The following simple thought experiment illustrates the validity of the above-mentioned Wigner's work. It demonstrates that VMD is inconsistent with SR. Let S_A and S_B denote two sources of optical beams with positions at $\mathbf{r} = (\pm 1, 0, 0)$ (see Fig. 10.3). The figure is embedded in the (x,y) plane. The beams intersect at point O and each of them continues in its original direction. Namely, in the case of optical photons, no photon-photon scattering event takes place. Now let us examine this system from an inertial frame Σ' that moves nearly at the speed of light in the negative direction of the y-axis. In Σ', the photons of the beams are energetic. Therefore, if VMD is correct, then these energetic photons contain hadrons, and hadron-hadron scattering events should take place. This means that contradictory results are obtained from

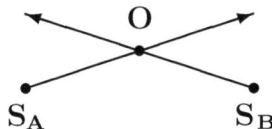

Figure 10.3: *Two beams of optical photons intersect at point* **O**.

two observations of the same process: One observer says that all photons continue to move in their original direction, whereas the other says that some photons are deflected to another direction. This contradiction proves that VMD is inconsistent with SR.

This discussion shows that the ad hoc modification (10.4) of the ordinary form of electrodynamics is unacceptable. The SM contains the electromagnetic and the strong interactions sectors. It follows that the SM provides no satisfactory explanation for hard photon-nucleon scattering that belongs to its domain of validity.

As of today, many textbooks ignore this problem, probably because the VMD idea violates SR in general and Wigner's work (see section 3.7 on p. 37) in particular. It means that ordinary students who have accomplished their physics studies are unaware of this intrinsic SM error. The textbook [94], which supports VMD, is an exception. This exception means that the book explicitly teaches an erroneous idea, whereas other textbooks ignore the problem. However, in either case, the inherent SM error of (10.4) is unsettled.

> *Conclusions: SM does not explain the hard photon-nucleon interaction. The VMD idea, which aims to settle this problem, is inherently wrong. The removal of this topic from most textbooks certainly does not solve the problem. The RCMT explains this effect.*

10.5.3   The Nuclear Force

The distance dependence of the nuclear potential takes the following form: At a short distance, it has a strong repulsive force that decreases rapidly; outside the repulsive region, it has an attractive component. The attractive component decreases much faster

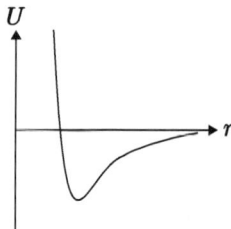

Figure 10.4: *Shape of the distance dependence of the nuclear potential U.*

than the $1/r$ Coulomb potential does. The graph describing the distance dependence of the nuclear potential is shown in Fig. 10.4. The original figure is depicted on p. 97 of [8].

It turns out that the graph describing the potential found between neutral molecules (or atoms of a noble gas) and the nuclear potential takes the same shape (see [95], p. 16). (This molecular force is also called the van der Waals force.)

The close similarity between the shape of the nuclear and molecular potentials suggests that these interactions rely on similar theories and that the atomic structure is similar to the structure of nucleons. This physical evidence is clearly in strong support of the RCMT claims of RCMT.1 - RCMT.3 of subsection 10.3.4 on p. 126.

The problematic issues that nuclear data present for QCD were recently admitted by Wilczek, who is one of the contributors to the QCD theoretical structure [96]. He stated, "...the original problem of understanding nuclear forces has rather fallen by the wayside." Wilczek continued, "Ironically, from the perspective of QCD, the foundations of nuclear physics appear distinctly unsound." According to Wilczek, some steps towards a solution to these QCD problems have been taken via numerical calculations that use lattice QCD algorithms [97]. The authors of [97] claimed that their calculations reproduce the potential graph of Fig. 10.4.

As admitted in [96], the calculations of [97] cannot be regarded as the final word. Two aspects of the questionable status of [97] are mentioned here. First, the authors of [97] use a pion mass $M_\pi \simeq 0.53$ GeV, which is about four times larger than the pionic physical mass. Another issue is their use of a Yukawa interaction mediated by a pion and similarly for vector mesons that account for the repulsive force. The Yukawa analysis relies on a boson wave function of the form $\phi(x^\mu)$, which depends on a single set of four space-time coordinates x^μ. It follows that the Yukawa function $\phi(x^\mu)$ describes a point-like structureless elementary particle. However, the pion is certainly a different object: It is not an elementary particle but a quark-antiquark bound state and its wave function depends on a greater number of independent variables $\Phi(\boldsymbol{x}_1, \boldsymbol{x}_2, t)$. Moreover, it is not a point-like particle because its mean square charge radius is not much smaller than that of the proton [29]. Hence, an application of mesons – which are quark-antiquark bound states – as carriers of attractive and repulsive interactions between particles is merely a *phenomenological* approach that may be useful for practical purposes but *certainly cannot be regarded as a basis for substantiating a theory like QCD.*

The Google citation utility shows that as of February 28, 2021, the paper [97] had been cited 535 times. This is a relatively large number of citations, indicating that many particle physics authors (and their referees) are still unaware of the distinction between elementary point-like particles and composite particles like the π and the ρ mesons. Another remark on this matter is as follows: Nuclear physics represents well-established experimental data, whereas QCD is the SM's strong interaction theory. Hence, the Wilczek quotation given above should be restated: *"Ironically, from the perspective of the real world, the foundations of QCD appear distinctly unsound."*

> Conclusion: QCD is about 50 years old, but it still does not explain a fundamental phenomenon like the nuclear force. The RCMT properties of RCMT.3 of subsection 10.3.4 provide an automatic explanation for this force.

10.5.4 A Further Examination of Nuclear Data

Let us examine data of some light nuclei.

N.1 The deuteron is a proton-neutron nucleus, and its binding energy takes the relatively small value of 2.2 MeV. (The binding energy of a typical nucleus is about 8 MeV per nucleon.) The valence quarks of the proton and the neutron are uud and udd, respectively. Hence, the deuteron comprises $uuuddd$ quarks. According to the three colors of QCD, the Pauli exclusion principle allows this system to be in a *strongly* bound state, where the binding energy is several hundred MeV. The true situation, where the deuteron's binding energy is just 2.2 MeV, categorically refute QCD.

Table 10.3: Nuclear Binding Energy (BE) and Radius

Name	Nucleons	BE	BE/Nucleon	Radius
Deuteron	pn	2.2	1.1	2.14
^{3}He	ppn	7.72	2.57	1.97
Triton	pnn	8.48	2.83	1.76
^{4}He	ppnn	28.3	7.08	1.68

N.2 Table 10.3 provides further arguments that support the previous point. It proves that for very light nuclei, the BE/nucleon increases with the number of nucleons, whereas the radius of these nuclei decreases. This evidence negates the QCD color assumption. The data show that the BE per nucleon of the deuteron is much smaller than that of the other nuclei of this table. According to the QCD color assumption, the Pauli exclusion principle allows the deuteron's *uuuddd* quarks to be in a strongly bound state and pushes some quarks of the larger nuclei to a higher energy level. Hence, QCD expects that the binding energy per nucleon of the deuteron *will not be considerably smaller* than that of the other nuclei. This QCD expectation is inconsistent with the data.

N.3 Consider the four $\Delta(1232)$ baryons of section 10.2, on p. 119. All these baryons have about the same mass, and the Δ^{++} comprises three *uuu* quarks. The QCD color degree of freedom says that it must be in the lowest quark state, where the quarks are in the single-particle s-wave. However, this same single-particle s-wave is ascribed to the proton, which comprises *uud* quarks. Hence, if QCD is right, the mass of the Δ^{++} should be about the same as that of the proton. The data disprove this QCD attribute, and the mass difference between these particles is about 300 MeV.

> Conclusion: Well-established data of light nuclei and baryons prove that QCD is unsound (see also [96]).

10.5.5 The Nuclear Liquid Drop Model

More than 80 years ago, Weizacker suggested a model for describing nuclear mass. His successful model relies on nuclear data, where the "nuclear binding energy is based on the analogy of a nucleus with a drop of incompressible fluid. We have seen earlier that nuclear volume increases linearly with the number of nucleons, in support of such a 'liquid drop' model" (see [8], p. 139). Here we have a description of two kinds of data – the nuclear radius and the nuclear binding energy. Like in an incompressible liquid drop, and excluding few very light nuclei, the nuclear volume is (very nearly) proportional to the number of nucleons. Similarly, the nuclear binding energy is analogous to that of a liquid drop.

This feature is inconsistent with QCD. For example, Wilczek [96] explains: "But why don't the separate proton and neutron bags in a complex nucleus merge into one common bag? On the face of it, the one-bag arrangement has a lot going for it. It would allow quarks and gluons free access to a larger region of space, and so save on the energetic cost of localizing their quantum-mechanical wavefunctions. But in such a merger, protons and neutrons would lose their individual identities, and our traditional, quite successful model of atomic nuclei would crumble. What prevents that calamity?"

The RCMT explanation of the similarity between neutral atoms and nucleons (see item RCMT.3 on p. 126) also explains the successful analogy between ordinary liquids and the nuclear liquid drop features.

> Conclusion: Properties described by item RCMT.3 of subsection 10.3.4 provide the RCMT's automatic explanation for the data of the radius of nuclei and for their binding energy. Referring to QCD, one finds again the relevance of Wilczek's description [96]: "Ironically, from the perspective of QCD, the foundations of nuclear physics appear distinctly unsound."

10.5.6   The Nuclear Tensor Force

Another nuclear phenomenon is the nuclear tensor force. The existence of this force is inferred from the deuteron's prolate shape (see [8], p. 65). This kind of force and its sign are consistent with electromagnetic-like interaction between two particles with the same dipole sign relative to their spin direction [28], p. 143):

$$V_{Dipole} = -[3(\boldsymbol{\mu}_1 \cdot \boldsymbol{r})(\boldsymbol{\mu}_2 \cdot \boldsymbol{r}) - r^2 \boldsymbol{\mu}_1 \cdot \boldsymbol{\mu}_2]/r^5. \qquad (10.5)$$

The dipole-dipole interaction (10.5) (also called the tensor force) proves that the nuclear tensor force is not related to the magnetic moments of the proton and the neutron because these magnetic moments have opposite signs. The RCMT provides a satisfactory explanation for the origin of the nuclear tensor force. Indeed, the nucleons are spin-1/2 particles where the state is characterized by three u and d quarks of the same isospin structure. Analogously to electrodynamics, spinning monopoles should create an *axial* electric dipole moment, and isospin symmetry confirms that the proton and the neutron have the same size and sign of an *axial* electric dipole moment, as required by the deuteron data.

The *axial* electric dipole moment that RCMT assigns to nucleons is consistent with the experimentally known vanishing value of the neutron *polar* electric dipole moment. Indeed, the fields of the polar dipoles are bound and not radiation fields. Furthermore, measurements of the neutron's electric dipole moment use ordinary electromagnetic devices that are affected by electrons. Using the RCMT result of RCMT.2 of subsection 10.3.4 on p. 126, one realizes that these devices are blind to monopole bound fields of *axial* electric dipoles. Therefore, the measurements showing a vanishing electric dipole moment of the neutron just verify the null value of the neutron's *polar* electric dipole moment and that the neutron's state is determined by parity conserving monopole interaction.

A common property of a Dirac-like monopole theory is that electromagnetic fields of a charge are identical to those of a monopole. The foregoing discussion also explains why the null result of the measured electric dipole moment of the neutron proves that a Dirac-like monopole theory is unsuitable for explaining strong interactions.

> Conclusion: The RCMT explains the nuclear tensor force, as well as the null result of the measured electric dipole moment of the neutron. A Dirac-like monopole theory fails to do that. QCD does not address the nuclear tensor force.

10.5.7 The EMC Effect

Another nuclear effect is the variation of the volume occupied by nucleonic quarks as a function of the number of nucleons in nuclei. A report of this quark distribution has been published [98], and its results were confirmed soon after [99]. The outcome of these experiments is known as the EMC effect. This effect shows that *the nucleon's quark volume increases with the increase of the nucleon number in nuclei.* Before the EMC experiment, QCD supporters published an erroneous prediction for the EMC effect [98]. However, the RCMT hadronic theory provides a straightforward explanation for this effect. Indeed, atomic-like screening effects of Maxwellian electrodynamics also hold for RCMT monopoles – namely, quarks – of nucleons. Hence, in a nucleus, quarks of a given nucleon penetrate neighboring nucleons. The penetrating quarks screen the field of the core of the neighboring nucleon, and the volume of its own quarks increases. Evidently, the effect increases with the average number of neighboring nucleons, which means that the effect is larger for heavier nuclei. An analogous effect exists in solids and liquids [100]. In contrast, QCD supporters admit that they have still not provided an adequate explanation for the EMC effect [101], although it was discovered several decades ago.

> Conclusion: The EMC effect shows that the self-volume of quarks in a nucleus increases with the number of nucleons. The RCMT argues that due to the charge-monopole dual relations, the screening effect of atomic electrons also holds for nucleons' quarks. Therefore, the EMC effect can be regarded as another support for the RCMT as a theoretical basis for strong interactions. In contrast, QCD supporters have provided an erroneous prediction for the EMC effect, and even after several decades, they still have no explanation for it.

10.5.8 The Significance of Hadronic Mass

Table 10.4 shows the mass (in MeV) of four hadrons and their valence quarks. The fourth column shows the mass difference between the two baryons and the two mesons. The number in the last column is the appropriate difference between the numbers in the fourth column. Here, π^+ is the lightest positively charged meson and K^+ is the lightest positively charged strange meson. Similarly, the proton is the lightest baryon and the Σ^+ is the lightest positively charged hyperon.

Table 10.4: Mass Data of Four Positively Charged Hadrons.

Particle	Quarks	Mass	Difference	Δ
Σ^+	uus	1189	251	-
Proton	uud	938	-	-
K^+	$u\bar{s}$	494	354	103
π^+	$u\bar{d}$	140	-	-

The deuteron D is a clear example of the concept of binding energy BE of particles (see [8], p. 10)

$$1876 = M(D) = M(p) + M(n) - BE(p+n), \qquad (10.6)$$

One can write analogous expressions for the particles of table 10.4

$$1189 = M(\Sigma^+) = M(uu) + M(s) - BE(uu + s); \quad (10.7)$$
$$938 = M(p) = M(uu) + M(d) - BE(uu + d); \quad (10.8)$$
$$494 = M(K^+) = M(K^-) = M(s) + M(\bar{u}) - BE(s + \bar{u}); \quad (10.9)$$
$$140 = M(\pi^+) = M(\pi^-) = M(d) + M(\bar{u}) - BE(d + \bar{u}). (10.10)$$

Assuming that the Σ^+ uu quark structure of (10.7) is about the same as that of the proton of (10.8), one subtracts (10.8) from (10.7)

$$251 = M(\Sigma^+) - M(p) = M(s) - M(d) + BE(uu+d) - BE(uu+s). \tag{10.11}$$

Similarly, (10.9) and (10.10) yield

$$354 = M(K^-) - M(\pi^-) = M(s) - M(d) + BE(d+\bar{u}) - BE(s+\bar{u}). \tag{10.12}$$

A subtraction of (10.11) from (10.12) yields

$$103 = [BE(d+\bar{u}) - BE(s+\bar{u})] - [BE(uu+d) - BE(uu+s)]. \tag{10.13}$$

This outcome proves an interesting result:

> A comparison between d and s quarks proves that in baryons the s quark is bound stronger than the d quark. In meson the opposite result holds. In contrast, the small mass of π mesons proves that in mesons, the binding energy of the u and d quarks is much greater than their binding energy in the proton.

The RCMT easily explains the effect described above. Baryons contain inner closed shells of u and d quarks, and quarks are ordinary Dirac particles. Hence, because of the Pauli exclusion principle, the binding energy of u and d quarks to the proton is smaller than that of the s quark. Mesons are $\bar{q}q$ bound states and the Pauli exclusion principle does not hold for them.

QCD does not discuss these data. It says that baryons comprise three valence quarks, gluons, and some $\bar{q}q$ pairs. For quarks, the QCD color degree of freedom increases the number of empty places by a factor of 3. Hence, QCD says that in the proton, the Pauli exclusion principle does not distinguish between one d or s quarks.

> Conclusion: Consequences of the data of Table 10.4 agree with the RCMT, whereas QCD has not provided an explanation for these data.

10.5.9 Outgoing Mesons

A well-known property of a baryon-meson interaction is the free disconnection between these particles. For example, the decay channel of a Δ baryon $\Delta \rightarrow n\pi$, where n denotes a nucleon, takes more than 99% of the events [29]. The Δ lifetime is of the order of 10^{-23} sec. This process means that a pion exits freely from a baryon's environment.

Let us examine qualitative aspects of how QCD and the RCMT treat the baryon-meson disconnection effect. The relevant points are as follows:

BM.1 Deep inelastic experiments prove that "the underlying process in electron-proton inelastic scattering is the elastic scattering of electrons from point-like spin-half constituent particles within the proton, namely the quarks" (see [79], p. 185)). Hence, at least effectively, quarks are point-like particles.

BM.2 As stated above, SM says that QFT is the fundamental theory of elementary particles. This theory depends on a Lagrangian density that has the form

$$\mathcal{L}(\psi(x), \psi(x)_{,\mu}). \tag{4.5}$$

Here $x \equiv (t, \boldsymbol{x})$ denotes the four space-time coordinates. The general community agrees on this issue. For example, "all field theories used in current theories of elementary particles have Lagrangians of this form" (see [20], p. 300).

It is explained in subsection 4.3.1 on p. 46 of this book why $\psi(x)$ describes a point-like particle. Hence, item BM.1 shows that such a function properly describes quarks.

An important feature of quantum theories is that the interaction is calculated as an integral over the particle's spatial points. This principle is the basis for the successful calculations of electronic states and energies of atoms and the explanation of the deep inelastic electron-proton scattering.

BM.3 Consider a Δ decay where the outgoing meson is π^+. The $u, \bar{d}$ quarks compose the π^+. The data prove that the proton's RMS charge radius is 0.84 fm and the π^+ RMS charge radius is 0.66 fm [29]. The locality of the interaction of the point-like quarks and the relatively large volume of the baryon and meson means that *the overall baryon-meson interaction is the superposition of the local interaction of the individual quarks.*

The RCMT provides a straightforward explanation for the effect. A meson is a $\bar{q}q$ bound state because its $\bar{q}q$ quarks carry a positive and a negative unit of monopole strength. Hence, due to the electrostatic-like attraction between opposite monopoles, quarks of the pion are strongly bound. In contrast, the pion's (and any other meson's) overall monopole strength vanishes. Therefore, it is not affected by any electrostatic-like monopole attraction of the baryonic environment (except for a residual nuclear-like force). In contrast, according to QCD, both antiquarks and quarks are attracted by the other quarks so strongly that the effect of confinement occurs in nucleons and mesons. QFT locality implies that this attraction should confine the u quark of the π^+ meson and the $\bar{d}$ antiquark of this meson.

Remark: QCD states that "the nonobservation of free quarks is explained by the hypothesis of color confinement, which states that colored objects are always confined to color singlet states" (see [79], p. 248). Several reasons illustrate why this hypothesis is irrelevant to the problem discussed here:

R.1 It is a hypothesis and not a dynamic argument.

R.2 By imposing a new law of physics, this hypothesis explains why a colored quark combination cannot exist as a free particle. Hence, it does not apply to the dynamic laws of a white quark combination.

R.3 Following from the previous point, QCD says that a colorless pion *may* exit freely from the baryonic environment. This argument does not negate the dynamic arguments presented here. Indeed, these arguments prove that if QCD is right, although its white pion *may* exit freely from the baryonic environment, dynamic QFT laws actually prove that it *cannot* exit from this environment.

If QCD is correct, then QFT locality implies that confinement should apply to mesons in baryons and not only to quarks in nucleons or antiquarks in mesons. Therefore, it is not clear how QCD can explain the free exit of pions from the baryonic region.

> Conclusion: The RCMT provides a straightforward explanation for the well-known baryon-meson free disintegration effect. The additional QCD color hypothesis explains why a white pion *may* exit a baryon environment. In contrast, QCD textbooks do not explain why QFT self-evident dynamic laws, together with QCD confinement, make a situation where a pion *cannot* exit a baryon environment.

Remark: According to the RCMT, a baryon-meson interaction is analogous to the nuclear force, which is a residual force between particles where the overall monopole strength vanishes. Hence, the practically free baryon-meson disintegration is a nuclear-like phenomenon. This is another example of the relevance of the statement of section 10.5.3 on p. 132: "Ironically, from the perspective of the real world, the foundations of QCD appear distinctly unsound."

10.5.10 Quark Confinement

Experimental evidence proves that "quarks are *absolutely confined* within baryons and mesons, so that no matter how hard you try, you cannot get them out" (see [67], p. 42). This means that the people who have constructed QCD faced *two* problems: "But it had two conspicuous defects: the experimental absence of free quarks and inconsistency with the Pauli principle" (see [67], p. 44). In contrast, this book shows that the quark's problem with the Pauli exclusion principle stems from a wrong interpretation of the data (see section 10.2, on p. 119). This argument, as well as the many QCD experimental failures that are pointed out in this chapter, explain why QCD should be rejected. This conclusion means that the QCD confinement hypothesis does not hold, and the quark confinement effect needs a coherent interpretation.

It turns out that one does not need to go too far to find a meaningful explanation of the quark's confinement. Indeed, this explanation can readily be found in a textbook stating, "A possible scenario for quark confinement: as we pull a u quark out of the proton, a pair of quarks is created, and instead of a free quark, we are left with a pion and a neutron" (see [67], p. 72). Furthermore, the data accumulated during recent decades show that there is no need to *create a $\bar{q}q$ pair,* because the additional pairs are explicitly found in the proton (see subsection 10.5.14 on p. 147).

The RCMT provides a straightforward explanation for this scenario. The strong electrostatic-like attraction of the quark's monopole to the baryonic core means that a huge amount of energy is required to remove it. In contrast, a meson (namely, a $\bar{q}q$ pair) is neutral respecting monopoles. Hence, as shown in section 10.5.9 on p. 139, it exits freely from the baryonic region.

10.5.11 Proton-Proton Cross-Section I

The total electron-proton cross-section is well documented in textbooks. At high energy, an inelastic process stems from an elastic

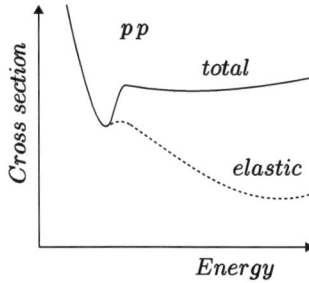

Figure 10.5: *Energy dependence of the total and the elastic proton-proton cross-section.*

scattering of an electron that hits an individual quark of the proton (see [79], p. 185). General QED formulas show that the electron-proton cross-section decreases monotonically with the increase of the collision energy (see [79], chapter 8.2). Moreover, "because of the finite size of the proton, the cross-section for electron-proton elastic scattering decreases rapidly with energy.   Consequently, high-energy $e - p$ interactions are dominated by inelastic scattering processes where the proton breaks up" (see [79], p. 178).

The corresponding data of high-energy proton-proton scattering have been known for several decades (see p. 11 of [102]). Figure 10.5 depicts the main features of these data. (The next subsection discusses recent data.) The proton-proton high-energy data differ dramatically from the corresponding electron-proton data: In electron-proton scattering, the total cross-section decreases monotonically, and the relative portion of the elastic scattering becomes negligible. In contrast, with high-energy scattering, the proton-proton total and elastic cross-sections begin to rise, and the relative portion of the elastic cross-section takes a uniform value of about 1/6. QCD has no explanation for this effect. For example, QCD says that a pp collision is a superposition of individual quark-quark collisions. Hence, it is not clear why a heavy blow of an electron on a quark yields an inelastic process where the proton breaks up, while in the case of a quark-quark heavy blow, the relative portion of elastic events is not negligible. QCD textbooks ignore this effect, and this negligence substantiates this assertion. On top of that, it is important to restate that a higher energy process is determined at a shorter distance: "To probe *small distances* you need *high energies*" (see [67], p. 6). Hence, in sheer contrast to the data, the QCD asymptotic freedom entails that the decrease of the proton-proton cross-section should be faster than the decrease of the electron-proton cross-section.

The RCMT provides a straightforward explanation for the data given above. At higher energy, more details of the proton show up, and its core gradually begins to participate in the scattering process. Because of its closed shells of quarks, the core is a relatively rigid object. Hence, a core collision in general and a core-core collision in particular, are likely to yield an elastic event. Moreover, the gradual increase of the contribution of the nucleon's core to the scattering process explains the increase of the total and the elastic cross-section.

> Conclusion: The RCMT core and its closed shells of u and d quarks explain why the elastic and total cross-section data of the proton-proton scattering stop decreasing. These data strongly contradict the QCD asymptotic freedom, and the QCD literature ignores this effect.

10.5.12   Proton-Proton Cross-Section II

The previous section examined proton-proton scattering data that have been known for more than three decades (see [23], p. 134). These data clearly refute QCD and its quite weird asymptotic freedom attribute. Experimental physicists continue to work, and the TOTEM collaboration of the LHC recently reported data of higher energy proton-proton scattering [103]. It is interesting to see how the total and the elastic proton-proton cross-sections behave at higher energy (see Fig. 10.6). The main points of the new data are as follows:

1. The total cross-section *continues to rise* with the increase of collision energy.

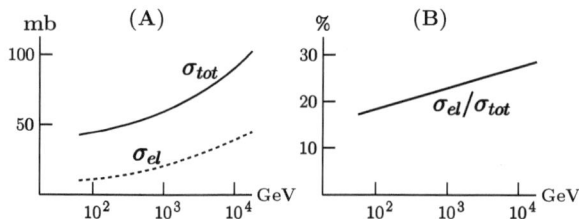

Figure 10.6: *Description of TOTEM's recent data: Energy dependence* ($\sqrt{S}$) *of: (A) The total and the elastic pp cross-sections; (B) The ratio between the elastic and the total pp cross-sections.*

2. The elastic cross-section *continues to rise* with the increase of collision energy.

3. The ratio between the elastic cross-section and the total cross-section *increases* with the increase of collision energy.

These data worsen the problematic status of QCD as explained in the previous subsection. In contrast, the rising role of the RCMT's core explains these data.

> Conclusions: The new data of the TOTEM collaboration on high energy pp scattering provide further support for the RCMT and even stronger reasons for the QCD refutation. In particular, QCD's asymptotic freedom, which is a vital element of this theory, says that at high energies, pp cross-section should decrease faster than that of the ordinary electron-proton scattering. The data decisively refutes this QCD element.

10.5.13   Pion-Proton Cross-Section

The PDG shows the $\pi^{\pm}p$ elastic and total cross-section [29]. Let us examine the RCMT and QCD treatments of the high-energy pion-proton elastic cross-section and compare it to the proton-proton elastic cross-section. A common experimental element used by these theories is that because of the pion's small mass, its $\bar{q}q$ components are more tightly bound than are the proton's valence quarks. Another self-evident element says that rigid components of the colliding particles make a dominant contribution to the elastic process at the energy region where the cross-section stops decreasing. Furthermore, for collision energy where the elastic cross-section stops decreasing, the rigid component is enclosed in a small spatial region. Indeed, as stated above, "to probe *small distances* you need *high energies*" (see [67], p. 6).

The RCMT says that the proton has a core that comprises inner closed shells of u and d quarks. These closed shells are analogous to the inner closed shells of atomic electrons. Hence, they are a rigid element of the proton, and this rigidity is the origin of the portion of the high-energy elastic pp scattering. Although the pion's quarks are bound more tightly than the proton's valence quarks are, the rigidity of the pion is certainly weaker than that of the proton's closed shells. Moreover, the rigidity of the pion is a property of the entire particle and not of a small spatial element of this particle. Therefore, the RCMT says that for energies where the pp cross-section stops decreasing, the process involves two rigid cores, while in the pion-proton collision, there is only one rigid core. For this reason, this theory says that in the high energy experiments discussed herein, the elastic pion-proton cross-section should be smaller than that of the pp cross-section.

QCD denies the existence of closed shells of quarks inside the proton. Thus, because of the pion's tightly bound $\bar{q}q$ components, it expects that the elastic pion-proton cross-section will not be smaller than the elastic proton-proton cross-section.

Let us examine the high-energy cross-section data for energies that are higher than the point where the elastic cross-section stops decreasing. Figures 10.7 and 10.8 show the data of the proton-proton and the $\pi^+ p$ cross-sections, respectively. The figures show that at this point, the proton-proton elastic cross-section is about 7 mb. In contrast, the corresponding value of the $\pi^+ p$ cross-section is about 3 mb. This outcome supports the RCMT and disproves QCD.

Conclusions: Previous subsections proved that QCD totally fails to explain the pp elastic and total scattering data. Hence, one should not be surprised by the information described here, which shows the QCD failure to explain the high-energy $\pi^+ p$ elastic data. By the same token, one realizes the success of the RCMT.

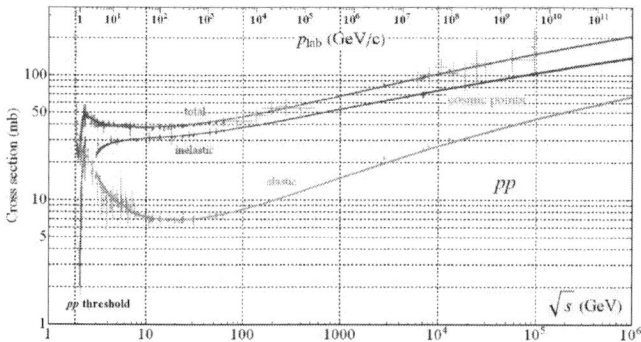

Figure 10.7: *Energy dependence of the elastic, inelastic, and total pp cross-sections (see the PDG report [29]).*

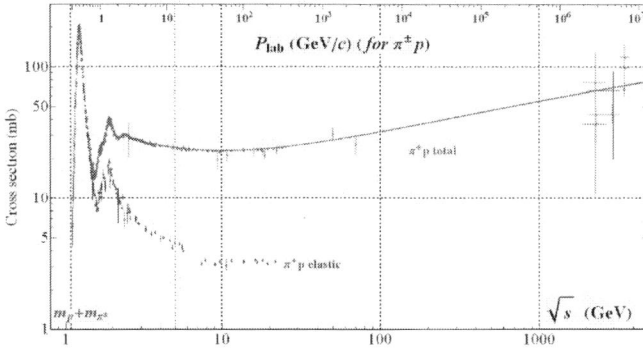

Figure 10.8: *Energy dependence of the elastic and the total $\pi^+ p$ cross-section (see the PDG report [29]).*

10.5.14 Antiquark Components of the Proton

Measurements show the momentum distribution of the nucleon's quark and its antiquarks as a function of Bjorken x ([23], p. 281; [79], p. 202). Figure 10.9 depicts the main features of these data. The nucleon's quarks are confined inside its volume, and the Heisenberg uncertainty principle proves that they acquire a Fermi motion. This motion is the underlying reason for the width of the graphs of Fig. 10.9 (see [23], pp. 266-271). In other words, the stronger the Fermi motion of a given component of the nucleon is, the wider its graph becomes in Fig. 10.9.

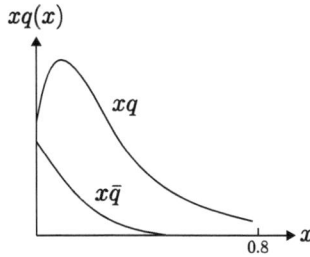

Figure 10.9: *Bjorken x dependence of the momentum distribution of the nucleon's quarks and its antiquarks.*

The data depicted in Fig. 10.9 show that the graph of the nucleon's quarks is much wider than that of its antiquarks. This means that the quarks' Fermi motion is considerably stronger than that of the antiquarks. Therefore, the Heisenberg uncertainty principle says that the nucleon's quarks are enclosed inside a much smaller volume than that of its antiquarks.

The charge radius of the proton is $<R_p>= 0.84$ fm, whereas that of the π^+ is $<R_{\pi^+}>= 0.66$ fm [29]. These data are relevant to the following argument: The π^+ is a $u\bar{d}$ bound state, and its radius is smaller than that of the proton. Now, QCD says that the proton comprises three valence quarks, a probability of quark-antiquark pairs, and gluons. Consider a state of one additional quark-antiquark pair. Here, QCD cannot explain why the proton's four quarks (the three valence quarks and the antiquark's companion) cannot hold the antiquark inside their self-volume while the single u quark of the π^+ holds the antiquark firmly inside a smaller volume.

The RCMT provides a straightforward explanation for the nucleon's larger spatial volume of antiquarks. The nucleon has a core with a monopole strength of $+3$. Quarks have a monopole strength of -1. Therefore, electromagnetic-like laws state that the core attracts quarks. The proton's valence quarks do not completely screen the core's monopole field at the proton's volume. Therefore, antiquarks that have a monopole strength of $+1$ are repelled to the nucleon's peripheral region.

Conclusion: The data show that antiquarks are pushed to the proton's peripheral region. QCD textbooks ignore this effect. Hence, QCD has no explanation for it. In contrast, the RCMT provides a straightforward explanation for this effect.

10.5.15 The $\bar{s}s$ Quark Pair in the Proton

A recent publication by the CERN ATLAS collaboration shows that the number of $\bar{s}s$ quark-antiquark pairs in the proton is about the same as that of the $\bar{d}d$ quark-antiquark pairs [61]. This publication also shows that the probability of finding one additional quark-antiquark pair is about unity. It is not clear how QCD can explain these new data. Indeed, a general rule of physics says that the probability of a higher energy state is smaller than that of a lower energy state. Here, the s-quark is heavier than the d-quark, and this explains why strange hadrons are heavier than their corresponding hadrons that comprise only quarks of the u and d flavor. For example, the mass of K^+ is 494 MeV, whereas the mass of π^+ is 140 MeV. Furthermore, the neutron's udd valence quarks indicate that the Pauli exclusion principle allows the addition of a $\bar{d}d$ pair to the proton's ground state. In contrast to QCD, the RCMT provides a direct explanation for these results. Indeed, the RCMT shows that in addition to the three valence quarks, the proton has a core that comprises an inner object *and* closed shells of u and d quarks. Furthermore, the RCMT treats quarks as ordinary Dirac particles (see item RCMT.3 of subsection 10.3.4). Hence, due to the baryon's u and d closed shells, the u and d quarks of an additional quark-antiquark pair occupy a higher energy state. In contrast, energy considerations indicate that in the proton, the existence of closed shells of the s quark is less likely. Hence, the Pauli exclusion principle allows the s quark of the additional $\bar{s}s$ pair to occupy a lower energy state. This effect increases the probability of the additional $\bar{s}s$ pairs. It can be concluded that two effects practically cancel each other and yield a similar probability for the additional $\bar{s}s$ and $\bar{d}d$ pairs: The higher mass of the s quark reduces its probability, whereas the Pauli exclusion principle allows it to take a lower energy state and increases its probability.

It is interesting to note that the existence of closed shells of u and d quarks in the proton's core *and* the null (or much lower) probability of closed shells of s quarks was *predicted* by the RCMT: "The baryonic core contains closed shells of quarks of the u,d flavor" (see [40], p. 98). Moreover: "It follows that inner closed shells of s quarks either do not exist or that their number is smaller than those of the u, d quarks" (see [40], p. 110). This successful prediction provides another indication of the RCMT's veracity.

Conclusion: QCD has no explanation for the quite large probability of an $\bar{s}s$ pair of quarks in the proton. In contrast, the RCMT provides a straightforward explanation for this effect.

10.5.16 The $\bar{u}u$ and $\bar{d}d$ Quark Pairs in the Proton

The excess of the proton's $\bar{d}$ antiquarks over its $\bar{u}$ antiquarks has already been known for several decades [104]. The results of recent experiments that have been carried out at CERN and Fermilab support this effect [105]. A simple argument that explains this effect is as follows. The figure on p. 281 of [23] (Fig. 10.9 on page 147 illustrates its general features) proves that a negligible portion of the proton's antiquarks has Bjorken x greater than 0.3. The new data are shown on the left side of Fig. 2 in [105]. For the sake of the present discussion, it is enough to realize that at $x < 0.3$, the ratio $\bar{d}/\bar{u}$ varies between 1 and 1.75. It means that the proton comprises more $\bar{d}$ antiquarks than $\bar{u}$ antiquarks. An application of the QCD additional color degree of freedom explains why QCD fails to account for this effect. In short, the QCD color degree of freedom allows adding one u quark and one or two d quarks to the proton's lowest energy state. Hence, QCD expects that the probability of the addition of one $\bar{u}u$ pair should be about the same as that of the addition of one $\bar{d}d$ pair. It follows that QCD is inconsistent with these data. It is interesting to note that the mainstream literature mentions QCD's failure to describe the different amounts of the $\bar{u}, \bar{d}$ components of the proton: "...but yet our understanding of the dynamics that form a physical proton from quarks and gluons is, at best, poor" [106]. In contrast, the Pauli exclusion principle and the proton's uud valence quarks are the reason for the straightforward explanation that is provided by the RCMT (see [40], p. 105). Here, quarks are ordinary Dirac particles. Hence, the proton's uud valence quarks and the Pauli exclusion principle mean that an addition of a d quark is energetically cheaper than an addition of a u quark.

> Conclusion: The excess of the $\bar{d}$ antiquark over the $\bar{u}$ antiquark in the proton components has been known for a long time. Further confirmation of this feature has recently been published. At present, QCD supporters explicitly state that their theory provides a poor description of the proton at best. In contrast, the RCMT provides a straightforward explanation for the excess of the $\bar{d}$ antiquark over the $\bar{u}$ antiquark in the proton components.

10.5.17 K^+ and K^- production in Proton Scattering

Experiments with protons have proved that besides the three uud valence quarks, the proton comprises $\bar{q}q$ pairs of the u, d, and s flavor [61]. Let us examine the near-threshold $K^\pm$ production in a proton-nucleus collision. The threshold is determined by the lightest overall mass of the outgoing particles that include $K^\pm$. The outcome of this collision is determined by strong interactions, and flavor is conserved. The threshold for K^+ production is determined by the mass of the particles of the following processes:

$$pp \to p\Lambda K^+; \quad pn \to n\Lambda K^+. \qquad (10.14)$$

Analogously, the threshold for K^- production is derived from

$$pp \to ppK^+K^-; \quad pn \to pnK^+K^-. \tag{10.15}$$

This process is a superposition of two primary processes – direct $s\bar{s}$ pair production that yields the K^+K^- production or ejection of an s quark of the proton's $s\bar{s}$ pair that already exists in the proton.

An examination of the cross-section just above the threshold provides important information. Here one should note that for a collision of elementary particles, the phase-space vanishes precisely at the threshold energy and it arises continuously for energies that are greater than this value. A similar situation holds in the collision of a proton with a nucleus. A process that belongs to (10.15) comprises four outgoing particles, whereas a process that belongs to (10.14) comprises three outgoing particles. Hence, just above the threshold, the phase-space of (10.15) grows faster than that of (10.14). Therefore, for energies that are just above the threshold, phase-space arguments are about the same for the K^- and K^+ production of the (10.15) process.

Experiments of kaon production by proton-nucleus scattering close to the production threshold show striking information (see fig. 1.a of [107]; fig. 1 of [108]). These experiments show that near the threshold, the number of K^+ production of the process (10.14) in proton-nucleus scattering is about 10 times larger than that of K^- near its threshold, which is produced by the process (10.15). This means that it is much easier to eject an $\bar{s}$ antiquark and produce a K^+ (namely, a $u\bar{s}$ bound state) than to eject an s quark and produce a K^- (namely, an $s\bar{u}$ bound state). Therefore, the data prove the effect delineated below.

> **Effect A:** *In the case of a proton collision with a nucleus near the threshold, it is much easier to eject an $\bar{s}$ antiquark and produce a K^+ meson than to eject an s quark and produce a K^- meson.*

The RCMT provides a straightforward explanation for this effect:

R.1 RCMT says that the structure of a baryon is analogous to that of an atom that has three electrons in its outer shell. Thus, a baryon has a core made of an inner object and some closed shells of u and d quarks. The overall monopole strength of this core is $+3$, and it is electrically neutral. A quark carries one negative unit of monopole strength (see section 10.3 on p. 122).

R.2 The monopole interactions are analogous to electromagnetic interactions. At the inner parts of a baryon, valence quarks do not completely screen the monopole field of the core's positive monopole strength. Hence, antiquarks that carry a positive unit of monopole strength are pushed towards the baryon's peripheral regions. In contrast, the quark of the $\bar{q}q$ pair is pulled towards the baryon's inner parts. Therefore, it is much easier to eject an $\bar{s}$ antiquark and produce a K^+ meson than to eject an s quark and produce a K^- meson.

This outcome shows how the RCMT explains Effect A.

QCD provides no explanation for the effect where antiquarks are pushed to the proton's peripheral region. Hence, one wonders how this theory will explain the effect discussed in this subsection.

> **Conclusion:** *Effect A of this subsection supports the RCMT.*

10.5.18 The Proton Spin Crisis

The present (August 2021) Wikipedia item on the proton spin crisis says: "The proton spin crisis (sometimes called the 'proton spin puzzle') is a theoretical crisis precipitated by an experiment in 1987 which tried to determine the spin configuration of the proton. The experiment was carried out by the European Muon Collaboration (EMC)" [109].

The item continues: "Physicists expected that the quarks carry all the proton spin. However, not only was the total proton spin carried by quarks far smaller than 100%, these results were consistent with almost zero (4-24%) proton spin being carried by quarks. This surprising and puzzling result was termed the 'proton spin crisis'."

This Wikipedia item summarized: "The problem is considered one of the important unsolved problems in physics."

The multiconfiguration feature of the states of atomic electrons, which is described in subsection 7.5 on p. 78, is relevant to the effect that has produced the proton spin crisis. While referring to this problem, one ought to stipulate that general properties of space require that a theoretical description of a *closed system* should provide a coherent expression for the *conserved total angular momentum J* of the entire system. Nothing can be said about the angular momentum of a specific particle that belongs to the system. The reason for this statement is simple: Particles that compose a system interact with one another, and their single-particle angular momentum is generally not a good quantum number. (For example, the earth and the moon rotate around their center of mass, and the earth-moon system rotates around the sun. Hence, considering the sun's frame, the angular momentum of the earth and the moon are not well defined.)

The multiconfiguration problem of the proton's quarks is even more complicated than the electronic problem of atoms. Entering details of this issue is far beyond the scope of this book (see, e.g., [9], [68]). Therefore, the arguments are described briefly:

MC.1 Experiments prove that strong interactions are much stronger compared with electromagnetic interactions. Therefore, spin-dependent interactions contribute a larger part to the overall state. In particular, the spin-orbit and spin-other-orbit interactions mix spin angular momentum with spatial angular momentum. Each configuration has a total spin value J and its z-projection J_z, which are equal to the J, J_z of the entire system. However, the single-particle spin components s, s_z of each configuration are not good quantum numbers.

MC.2 The existence of additional $\bar{q}q$ pairs of the u, d, s flavor, make a non-negligible contribution to the proton's spin structure (see subsection 10.5.15, 10.5.16), because a larger number of constituents produces a larger number of combinations of configurations that may contribute to the state.

For the reasons outlined above, many configurations are expected to contribute a non-negligible component to the proton's multiconfiguration state. Many of these configurations have a single particle spatial angular momentum that is greater than zero. A spin that is coupled to a non-zero spatial angular momentum may take either direction. This effect strongly reduces the quark's single-particle spin contribution to the proton state.

> Conclusion: The quite strange phenomenon where the entire particle physics community ignores the well-documented physical effect of the CI is the reason for the proton spin "crisis".

10.5.19 The Neutron's Mean Square Charge Radius

The neutron is an electrically neutral particle, but its mean-square charge radius takes a small *negative* value [29]. This subsection describes how the RCMT provides a qualitative explanation for this property of the neutron. The neutron's state is the isospin analog of the proton's state, and it is characterized by the *udd* valence quarks. It is shown below that two different effects push electrically negative components of the neutron to outer regions. These effects increase the negative value of the neutron's mean-square charge radius:

1. A fundamental element of the RCMT says that quarks are ordinary spin-1/2 Dirac particles. As discussed in [110] and by analogy to one of Hund's rules, the neutron's state favors spatially antisymmetric terms of the *dd* quarks. Evidently, an antisymmetric spatial state of two identical single-particle functions vanishes. Therefore, these states can be found in configurations that have a spatial excitation of one or two *d* quarks of the neutron. This means that the neutron comprises configurations where the *dd* quarks have a single-particle function with a larger mean radius. For this reason, the neutron's *d* quarks are more likely to be found in outer regions. Thus, because of the negative charge $-e/3$ of the *d* quark, the r^2 weight of the neutron's negative charge increases.

2. The neutron contains configurations with additional pairs of $\bar{q}q$ quarks. Using the proton's data from subsection 10.5.16 and the isospin symmetry, one finds that in the case of a neutron, a $\bar{u}u$ pair is more likely to be found compared with a $\bar{d}d$ pair. As shown in subsection 10.5.14, the volume of the nucleon's antiquarks is larger than that of quarks. The electric charge of the $\bar{u}$ quark is $-2e/3$, whereas that of the $\bar{d}$ quark is $e/3$. The overall arguments presented here boil down to the conclusion that the existence of antiquarks in the neutron increases the contribution to the negative value of the neutron's mean-square charge radius.

The points outlined above provide a qualitative explanation for the negative value of the neutron's mean-square charge radius and show that it is consistent with the RCMT foundation of the hadronic theory. Furthermore, if hypothetical experiments yielded a positive value for the neutron's mean-square charge radius, the RCMT hadronic theory would be disqualified.

> Conclusion: The RCMT hadronic theory explains the negative value of the neutron's mean-square charge radius. In contrast, QCD textbooks do not provide a theoretical explanation for this effect.

10.5.20 The Polarized Proton Scattering Experiments

An article by Krisch reported data on polarized proton-proton scattering. The results showed that at higher energy, a difference arises between the parallel spin data and the antiparallel spin data [111].

Krisch's description of the meaning of the results was as follows: "In particular, the theory that is now called QCD, has been unable to deal with this data: Glashow once called this experiment 'the thorn in the side of QCD'. In his summary talk at Blois 2005, Stan Brodsky called this result 'one of the unsolved mysteries of hadron physics'." Krisch continued, "Some theorists seemed quite unhappy" with the results of polarized proton experiments, and QCD experts have expected that "QCD might not work for elastic scattering." The biased and unscientific approach of QCD experts to this issue is inferred from Krisch's statement: "One result of our experiments was to make both elastic scattering experiments and spin experiments unpopular in some circles." Note that Krisch also mentions *the QCD unsettled problem with the proton-proton elastic scattering*. This problematic QCD topic is discussed above in subsections 10.5.11 on p. 142 and 10.5.12, on p. 144.

In principle, the RCMT explains this phenomenon. Indeed, the RCMT is an electromagnetic-like theory, and it is well known that in the electromagnetic case of electron collision, an energy increase entails an increase of the relative portion of the spin-dependent (magnetic) interaction over the spin-independent (electric) interaction (see [23], pp. 192-194). Hence, the RCMT asserts that in the case of higher energy experiments, the relative portion of the proton-proton spin-dependent interactions is likely to rise.

> Conclusion: People belonging to QCD circles adopt the unscientific policy of regarding experiments that refute their theory as "unpopular experiments." The electromagnetic-like RCMT explains the spin-dependent results of polarized proton-proton scattering.

10.5.21 The Strong CP Problem

Quantum textbooks discuss these transformations: The parity transformation (called P) of the space-time coordinates is $(t, \boldsymbol{x}) \to (t, -\boldsymbol{x})$. Charge conjugation (called C) is a transformation where a particle and its antiparticle are interchanged (see [14], p. 64). Experiments prove that strong and electromagnetic interactions conserve C and P (see [14], p. 64). Hence, these interactions conserve the combined CP transformation. These experimental data should affect the structure of theories that describe these interactions.

It turns out that the theoretical structure of QCD allows the CP violation. This well-known QCD property is called *the strong CP problem* (see [14], p. 726). This problem is regarded as one of the unsettled QCD problems [112].

The strong CP problem does not arise in the RCMT. Indeed, the RCMT is a regular Maxwellian-like theory, which is called a U(1) theory in the group theory parlance. In the quantum domain, it uses the Dirac field for a description of a massive particle. This theory conserves C and P (see [14], pp. 65-71). Hence, it conserves CP as well.

> Conclusion: The mathematical structure of QCD allows a CP violation. In contrast, experimental strong interaction effects that are relevant to QCD substantiate CP conservation. This is an unsettled QCD problem. Conversely, the RCMT is a Maxwellian-like theory that uses the Dirac equation for massive particles. This theory abides by C, P, and CP conservation. The experimental data support these properties.

10.5.22 The QCD Pentaquark Prediction

Language is a primary tool for communication between people. Consequently, people organize words in sentences that have the required meaning they wish to convey. In particular, a misunderstanding bears a negative effect on communication between scientists. Therefore, writers of a scientific text strive to use words and statements that have a well-defined sense. The need for clarity in a scientific text was put forward by Dirac in the following words: "In science you want to say something that nobody knew before, in words which everyone can understand. In poetry you are bound to say... something that everybody knows already in words that nobody can understand" [113].

The term *pentaquark* was coined in [114, 115]. In these articles, the authors used QCD considerations and defined a pentaquark as *a strongly bound state* of four quarks and one antiquark, where each of the five particles has a specific flavor. For example: "We restrict in the present paper to the third category which consists of multiquark states whose strong decay is strictly forbidden by energy, flavour and baryon number conservation" [114]. Similarly, "A new candidate for an exotic hadron is presented: an anticharmed strange baryon denoted by $P_{\bar{c}s}$, a bound state of a nucleon and an F (now called D_s)" [115]. An example of their idea is the following flavor configuration $uuds\bar{c}$: In general, these authors define a pentaquark as a strongly bound particle that comprises the quark flavor $uudQ_1, \bar{Q}_2$, where Q_1, Q_2 have a flavor that is heavier than that of the u and d quarks *and* Q_1, Q_2 do not have the same flavor. (Because of isospin symmetry, the configuration $uddQ_1, \bar{Q}_2$ is also a pentaquark.) It is clearly stated in [114, 115] that according to the QCD laws, a pentaquark is expected to be *bound strongly* and be energetically stable against decay into a baryon and a meson by strong interactions. These pentaquark attributes indicate how to detect pentaquarks in experiments. Hereafter, a particle that has the foregoing pentaquark properties is called *the original QCD pentaquark*.

> Definition: The *original QCD pentaquark* is a *strongly bound* particle that comprises $uudQ_1\bar{Q}_2$ or $uddQ_1\bar{Q}_2$ quarks, where Q_1 and Q_2 are quarks of a different flavor, each of which is heavier than the u and d quarks.

A feature of human language is that the meanings of some words change with time. One cannot be sure that scientific terminology is free of this effect. Two different examples where the original QCD pentaquark definition is extended are described below. They show that the present mainstream literature deviates from accurate scientific terminology. In particular, it does not adhere to the original QCD pentaquark definition and uses this term for different kinds of quantum systems. This means that an organization of pentaquarks in well-defined categories may contribute to the accuracy required from a scientific text. Evidently, pentaquarks that belong to the same category should have common physical properties and pentaquarks that belong to different categories should differ by at least one physically meaningful property. The main objectives of this subsection are the organization of pentaquarks in appropriate categories and the analysis of the meaning of the changes in the original QCD pentaquark definition.

Changes in the Meaning of Pentaquarks

Over time, the original meaning of the term pentaquark in QCD has been extended, and it now includes other kinds of quantum states. Let us substantiate this claim by showing two examples where the current SM literature assigns the term pentaquark to quantum states that do not belong to the original QCD pentaquark definition. These examples refer to the work of many authors whose articles have been published in mainstream journals. Therefore, they indicate that the general community has extended the meaning of the original QCD pentaquark definition. Unfortunately, the mainstream literature does not explicitly state that its definition of pentaquarks has been extended.

The first example refers to the possible existence of a particle called Θ^+ whose mass is about 1530 MeV. This particle was introduced more than 20 years ago, and it is discussed in the SM literature by many theoretical and experimental physicists. These discussions refer to Θ^+ using the term pentaquark. This particle comprises $uudd\bar{s}$ quarks [116]. Now, the neutron comprises udd valence quarks, and its mass is 939.6 MeV. Moreover, K^+ comprises $u\bar{s}$ quarks, and its mass is 493.7 MeV [29]. Hence, the relevant mass relation is

$$M(n) + M(K^+) = 1433.3 < 1530 = M(\Theta^+), \qquad (10.16)$$

where the numbers are in MeV. As stated above, the original definition of a QCD pentaquark says that *QCD considerations indicate that it is bound strongly.* In contrast, the above-mentioned mass data of (10.16) prove that Θ^+ is certainly an *unbound* state of the neutron and the K^+ meson because the sum of the mass of these particles is smaller than that of Θ^+. Because of the unbound state of Θ^+, the original QCD pentaquark definition means that it is not a genuine pentaquark. Incidentally, the existence of Θ^+ has no experimental support. Indeed, a recently published report states that "it is now generally accepted that there is no substantial evidence for the existence of the Θ^+ state" [117].

Another kind of extension of the pentaquark notion is shown in a report published by the CERN LHCb collaboration [118] and in earlier publications (see, e.g., [119]). This collaboration has recently announced a similar discovery [120]. These articles report the detection of two pentaquarks called $P_c(4380)^+$ and $P_c(4450)^+$. (The number enclosed in parentheses denotes the particle's mass in MeV.) The current SM literature contains many discussions of these results. The quark configuration of these pentaquarks is

$uudc\bar{c}$, and these pentaquark states are resonances of the proton and the J/ψ meson. Indeed, the proton is a uud quark state with a mass of 938.3 MeV, whereas the J/ψ meson is a $c\bar{c}$ quark state with a mass of 3096.9 MeV [29]. The sum of these mass values means that, as in Eq. (10.16), one finds that

$$M(p) + M(J/psi) = 4035.2 < 4380 = M(P_c(4380)), \qquad (10.17)$$

and the result is similar for the heavier P_c. Therefore, like the case of Θ^+, $P_c(4380)^+$ and $P_c(4450)^+$ are *unbound* states of the proton and the J/ψ meson. Furthermore, the quark configuration of these particles is $uudc\bar{c}$, which means that the additional $c\bar{c}$ quarks are of the *same flavor*. From either of these properties, one can conclude that the quantum states $P_c(4380)^+$ and $P_c(4450)^+$ do not fit the original QCD pentaquark definition.

The examples above show that the physical terminology that is commonly used at present does not abide by the original QCD pentaquark definition. Therefore, an organization of the pentaquark notion into well-defined categories is a timely assignment.

Categories of Pentaquarks

Below, the term pentaquark describes any configuration of four quarks and one antiquark. Every pentaquark mentioned here is either an experimentally confirmed quantum state or a theoretical particle of QFT or QCD. It is shown here that such pentaquarks can be found in many physical states. In the following, they are organized into four distinct categories where pentaquarks that belong to the same category have the same physically meaningful properties and each category is named appropriately:

1. Bound QFT Pentaquarks

   It is shown here that fundamental properties of QFT predict the existence of pentaquarks that are stable against a strong interaction decay. QFT describes states and processes where the number of particles may vary because of the existence of particle-antiparticle pair(s) where flavor is conserved (see e.g. p. 65 of [2]). The proton is a well-known stable particle where its structure has been extensively examined since the beginning of the accelerator era. The quark description of the proton generally takes the form of three components uud, called valence quarks. However, that the proton's state contains a significant probability of additional quark-antiquark pairs have already been recognized for many years (see [23],

p. 282). Pairs of the flavors u, d, s have been identified and analyzed (see subsections 10.5.15 on p. 148 and 10.5.16 on p. 149). Furthermore, flavor conservation of the strong and electromagnetic interactions proves that the form of these pairs is restricted to the combinations $\bar{u}u$, $\bar{d}d$, and $\bar{s}s$. For this reason, experimental evidence proves that the uud valence quarks do not represent the entire proton quark structure. Thus, one finds that the proton's quantum state is

$$\psi(p) = a_0\psi_0(uud) + a_u\psi_u(uuud\bar{u}) + a_d\psi_d(uudd\bar{d}) + a_s\psi_s(uuds\bar{s}) + \dots \quad (10.18)$$

Here, a_x denotes a numerical coefficient. It can be concluded that pentaquarks are found in elements that compose the proton, and because of its stability, the proton belongs to this pentaquark category. By the same arguments, the neutron is also included in this pentaquark category. Furthermore, there are other baryons like Λ, Σ, Ξ, and Ω that are stable with respect to strong interactions. The dynamic properties of these baryons are analogous to those of the proton. Therefore, the state of each of these baryons should have additional quark-antiquark pairs of the same flavor. Hence, these baryons also belong to this pentaquark category.

2. Unbound QFT Pentaquarks

Particles that belong to this category are unstable, and their quark structure is experimentally inferred from their decay products. Properties of three particles are discussed below to illustrate this issue.

The four $\Delta(1232)$ baryons $\Delta^{++}, \Delta^+, \Delta^0$, and Δ^- are examples of this kind of pentaquark. These particles are seen as a conspicuous broad resonance, and they decay into a nucleon and a pion (see p. 131 of [23]; [29]). Their πN decay mode proves that their structure has a component of the form $q_1q_2q_3q_4\bar{q}_5$, where each q_i is either a u or a d quark and the specific identity of these quarks is determined by the charge of the specific Δ baryon. For example, the Δ^{++} quark structure can be written as

$$\psi(\Delta^{++}) = a_0\psi_0(uuu) + a_u\psi_u(uuuu\bar{u})$$

$$+a_d\psi_d(uuud\bar{d}) + a_s\psi_s(uuus\bar{s}) + \dots \quad (10.19)$$

Relying on the mentioned decay products, one can conclude that each of the Δ baryons has pentaquark components. Because of their instability, the Δ baryons belong to this pentaquark category. Furthermore, the Δ^{++} is a baryon. Therefore, like in the case of the proton, QFT laws indicate that the last term of (10.19) should be a component in the description of Δ^{++}. For example, consider the quantum state

$$\psi_d(uuud\bar{d}) = \psi_p(uud) + \psi_{\pi^+}(u\bar{d}) \qquad (10.20)$$

of (10.19). Here, one sees the proton and the π^+ of the decay mode. A similar examination applies to every Δ baryon. Therefore, these baryons belong to this QFT pentaquark category.

Each member of the isospin doublet of the $N(1710)$ baryons (see [29]) is another example of a baryon that belongs to this pentaquark category. Consider the positively charged $N^+(1710)$ baryon. It has several decay modes, and one of them is the two-particle ΛK^+ channel. The quark structure of this decay is

$$N^+(1710) = uuds\bar{s} \rightarrow uds + u\bar{s} = \Lambda + K^+ \qquad (10.21)$$

The Λ baryon is characterized by the uds valence quarks, and its mass is 1115.7 MeV. The K^+ meson is characterized by the $u\bar{s}$ quarks, and its mass is 493.7 MeV [29]. One can use the arguments of (10.16) and (10.17), and write the expression

$$M(\Lambda) + M(K^+) = 1609.4 < 1710 = M(N(1710)). \qquad (10.22)$$

These values explain the instability of the N(1710) baryons with respect to the ΛK^+ decay channel. Relying on these decay products, one can conclude that the five quarks $uuds\bar{s}$ compose one of the quark configurations of the $N^+(1710)$ baryon. Because of its instability, $N^+(1710)$ belongs to this pentaquark category. Moreover, isospin symmetry proves that its isospin counterpart $N^0(1710)$ is also an analogous pentaquark and the quark configuration $udds\bar{s}$ is included in the description of $N^0(1710)$.

It turns out that the $P_c(4380)$ and the $P_c(4450)$ pentaquarks, which have recently been reported by the CERN LHCb collaboration (see subsection 10.5.22 on p. 158), are similar analogs of each of the two $N(1710)$ baryons. Indeed, relying on their decay modes, it was shown in the previous paragraph that each of the $N(1710)$ baryons contains the five quarks $uuds\bar{s}$ configuration. Similarly, as stated in the mentioned LHCb report, the decay modes of the $P_c(4380)$ and the $P_c(4450)$ pentaquarks show the existence of the five-quark $uudc\bar{c}$ configuration (see line 7 in [118]). Evidently, the $N(1710)$ baryon has already been known for several decades. Therefore, the two states that have been reported by the LHCb collaboration just refer to a replacement of the $s\bar{s}$ quarks of the $N(1710)$ by the $c\bar{c}$ quarks of the $P_c(4380)$ and the $P_c(4450)$. It is well known that pair production and pair annihilation are ordinary QFT processes [14]. Therefore, the replacement of $s\bar{s}$ by $c\bar{c}$ is nothing more than an ordinary QFT process. For this reason, the recent LHCb discovery certainly does not represent a new fundamental change in the pentaquark concept.

3. Lightly Bound Pentaquarks

Lightly bound pentaquarks comprise baryon-meson bound states where the binding energy is rather small and its strength is similar to the nuclear binding energy. The nuclear force is an example of a force that binds two nucleons. Similarly, a nuclear-like force may bind a nucleon and a meson, representing another example of two hadrons. Below, such a state is called a nuclear-like pentaquark. (The term *meson-baryon molecule* is used in the literature to describe this kind of pentaquark [121]. However, the term nuclear-like is used here because pentaquarks and nuclei comprise hadrons, whereas the molecular binding energy belongs to the physical properties of a system of electrons.) The nuclear force is known experimentally, and the mean nuclear binding energy per nucleon is less than 9 MeV. For this reason, the existence of this kind of pentaquark is independent of any specific physical theory, and the pentaquarks' binding energy should be a few MeV.

The arguments below indicate that the existence of nuclear-like pentaquark is highly questionable. Indeed, to be stable against a strong interaction decay, the pentaquark's meson should be in a spherically symmetric S-wave state because the excitation energy of other mesons is measured in hundreds of MeV. Indeed, a two-particle nuclear-like interaction cannot compensate for hundreds of MeV. Moreover, the state of an S-wave meson is similar to that of an atom of a noble gas. It is well known that noble gases show extremely low chemical reactivity [122]. These arguments mean that nuclear-like energy cannot compensate for the strong interaction energy of a meson where the spin does not vanish and a spin-0 meson is expected to behave like a noble gas. Therefore, finding a nuclear-like pentaquark is extremely unlikely. As of today, no such bound state of a pentaquark has been detected. This means that the experimental data are consistent with the prediction.

> *Remark: This book proves that weak interactions are spin-dependent processes. Hence, for high energy systems of heavy quarks, these interactions may produce a ground state meson where $j^\pi \neq 0^+$. In such a case the arguments presented here do not hold.*

4. The Original QCD Pentaquarks

As stated above, the term pentaquark was coined in 1987 [114, 115]. From the laws of QCD, these original works deduced that strongly bound hadronic states with a pentaquark structure should be found in accelerator data. Experimental efforts aiming to detect the original QCD pentaquarks, as well as theoretical discussions of this kind of particle, followed the publication of [114, 115].

QCD aims to describe the strongly bound state of the proton as a particle that comprises the three uud quarks, called *valence quarks*. Evidently, the flavor of every s, c, b quarks is not found among the proton's valence quarks. Therefore, according to QCD, an addition of quarks of the s, c, b flavor is not restricted by any direct or indirect version of the Pauli exclusion principle. The original QCD pentaquark proposal of [114, 115] follows this approach and clarifies that the quark configuration of this particle is $uuds\bar{Q}$, where Q denotes a heavy quark that is either c or b. This means that the existence of this state cannot be deduced from general QFT arguments that allow the addition of a $q\bar{q}$ pair of the *same flavor*. It is argued in [114, 115] that specific QCD properties provide appropriate binding energy that ensures the stability of the suggested pentaquark against strong decay.

The original QCD pentaquark could have been accidentally detected before 1987 or in the dedicated experiments that followed the publication of [118, 119]. As of today, there is no experimental confirmation of the existence of the original QCD pentaquark.

Conclusions

The pentaquark is a widely discussed subject pertaining to hadronic structure. For example, a current (August 2021) Google search shows more than 160,000 entries that contain the word pentaquark. The purpose of this subsection is to use the physical properties of pentaquarks to organize them in well-defined categories. To this end, experimentally confirmed stability and instability findings against strong interaction decay are used for the definition of pentaquark categories. The theoretical element of the analysis relies on the difference between general QFT properties and specific properties of QCD. Indeed, QFT is a general theory aiming to describe the strong, electromagnetic, and weak interactions that affect microscopic processes [14, 20]. In particular, pair production is a well-known QFT effect. In contrast, QCD is a specific theory that refers to hadronic structure and processes. The definition of the four pentaquark categories described above relies on these principles and the experimentally confirmed phenomenology of the nuclear force. The first and the second categories depend on the pair production effect of QFT. The third category depends on the well-known phenomenology of the nuclear force, and the last category depends on specific QCD properties.

This subsection shows that the first and the second pentaquark categories that rely on QFT prove that this general theory is successful in terms of experimentally confirmed pentaquark states. In contrast, despite the quite long pentaquark search, nuclear-like and QCD pentaquark states have not yet been detected in experiments. It is interesting to note that the failure of experimental attempts aiming to detect an original QCD pentaquark has been predicted by the author of this book [40].

The RCMT accepts the general laws of QFT. Therefore, pentaquarks that are based on pair production effects are consistent with the RCMT. Furthermore, other kinds of pentaquarks are nucleon-meson states. The overall monopole strength of each of these particles vanishes. Therefore, their interaction force is analogous to the residual force between nucleons.

> Conclusions: QCD experts have predicted the existence of a pentaquark that is a strongly bound particle comprising $uudQ_1\bar{Q}_2$ quarks, where Q_1, $\bar{Q}_2$ are heavier than the u and d quarks and Q_1, $\bar{Q}_2$ have a different flavor. Experimental efforts aiming to detect such a pentaquark have ended in vain. These failures can be regarded as another kind of QCD inconsistency.
>
> Contrary to QCD, the RCMT says that strongly bound pentaquarks, like those of the original QCD pentaquarks, cannot exist. The systematic failure of the search for the original QCD pentaquarks is another support for the RCMT.

A comparison with correct hadronic prediction clarifies the meaning of the systematic failure of attempts aiming to find strongly bound pentaquarks. In the early 1960s, Gell-Mann and Ne'eman independently predicted the existence of the Ω^- baryon. It turned out that Ω^- baryon was found about 3 years later [123], and the experimental technology of the time was much less efficient than that of later times. Furthermore, according to fundamental laws of physics, the antiparticle $\bar{\Omega}^+$ should exist, and indeed, this antiparticle has been found [29].

> *Conclusions: Correct predictions of hadronic states are found after a few years. In contrast, a systematic experimental failure that lasts for several decades casts serious doubt on the QCD's validity.*

10.5.23 The Strange Quark Matter Prediction of QCD

QCD supporters have argued that stable nuggets made of electrically neutral baryons, each of which resembles the $\Lambda(1116)$ baryon, should exist [124]. This kind of matter is called Strange Quark Matter (SQM). Many experimental attempts to detect SQM have been carried out for more than three decades. *All these experiments have ended in vain* [125].

In subsection 10.5.22, it was pointed out how the relatively underdeveloped technology of the early 1960s successfully found the correct prediction of the Ω^- baryon within a few years. This success is compared with the systematic failure to detect a pentaquark that supports the QCD prediction. This failure casts doubt on the QCD's veracity. The same argument applies to the systematic failure of the search for SQM.

Straightforward arguments of the RCMT say that SQM cannot exist: The lightest baryon that comprises *uds* quarks is the $\Lambda(1116)$ baryon [29]. Like any other baryon, it is neutral in terms of its monopole component. Hence, the intensity of its interaction with any other baryon is similar to that of the nuclear interaction, which is just a few MeV per nucleon. Evidently, a few MeV cannot compensate for the difference between the $\Lambda(1116)$ baryon and the nucleons, which is about 180 MeV. A fortiori, the state of any number of $\Lambda(1116)$ is extremely unstable against a weak decay. Hence, stable nuclear-like nuggets of $\Lambda(1116)$ baryons cannot exist. Experimental data consistently prove this prediction.

> QCD supporters have predicted the existence of nuggets of SQM. Many experimental searches for these objects have ended in vain. The RCMT says that SQM cannot exist.

Conclusion: The long-lasting futility of the experimental search for SQM supports the RCMT and casts doubt on the QCD's validity.

10.5.24   The QCD Glueball Prediction

A description of glueballs and their relevance to QCD is as follows: "One qualitative prediction of QCD is that there exist composite particles made solely of gluons called glueballs that have not yet been definitively observed experimentally. A definitive observation of a glueball with the properties predicted by QCD would strongly confirm the theory. In principle, if glueballs could be definitively ruled out, this would be a serious experimental blow to QCD. But, as of 2013, scientists are unable to confirm or deny the existence of glueballs definitively, despite the fact that particle accelerators have sufficient energy to generate them" [126]. In contrast to QCD, the RCMT says that gluons do not exist. A fortiori, glueballs do not exist.

The glueball prediction is another nearly 50-year-old QCD failure. Here, the remark on the fast detection of the Ω^- baryon, which is mentioned at the end of the pentaquark item (see subsection 10.5.22) is relevant. A prediction of a correct theory of particle existence is rapidly substantiated. Predictions of an incorrect theory are not supported by experiments.

> QCD supporters have predicted the existence of glue-
> balls. Many experimental searches for these objects
> have ended in vain. The RCMT says that glueballs
> cannot exist.

Conclusion: Glueball experiments support the RCMT and refute
the QCD.

10.5.25 The Three-Jet Event

A three-jet event is produced in a high-energy electron-positron
collision that yields three jets. A jet is a cluster of particles that
move in a direction that is enclosed within a narrow cone with its
tip at the collision point. It is assumed that each jet is produced by
a single particle created by the primary collision. QCD supporters
argue that these events prove the existence of the QCD's gluons
[127]. Thus, the collision creates a $\bar{q}q$ pair, and this pair emits a
gluon. Each of these three primary particles produces one jet. QCD
supporters say that "in 1979, experiments at the DESY laboratory
in Germany provided the first direct proof of the existence of gluons
– the carriers of the strong force that 'glue' quarks into protons,
neutrons and other particles known collectively as hadrons. This
discovery was a milestone in the history of particle physics, as it
helped establish the theory of the strong force, known as quantum
chromodynamics (QCD)" [128]. Several points illustrate why this
argument does not hold water:

3-J.1 The electromagnetic theory explains the bremsstrahlung effect in which colliding charged particles emit a photon (see [2],
section 7.6). It shows that for this process, the cross-section
of an electron-nucleus collision is proportional to $e^6 Z^2$, where
Z denotes the number of protons in the nucleus. The small
quantity $e^2 \simeq 1/137$ means that bremsstrahlung of colliding charges has a quite small effect in the electron-positron
collision (where Z=1) that yields the three-jet event. However, this argument does not hold for the RCMT and the
quarks that are involved in the hadrons that are produced
by the electron-positron collision event. Here, the quarks'
monopole strength is g, and $g^2 \gg e^2 \simeq 1/137$. Hence, the
bremsstrahlung analog of the quarks' monopole is expected
to yield a significant effect of a bremsstrahlung photon emission. The experimental evidence of the strong interaction of
a hard photon with a hadron explains the three-jet event.

3-J.2 QCD supporters report the theoretical examination of hard photons in their three-jet calculation. However, one wonders how can they do the required calculations coherently if the hadronic photon interaction has been swept under the carpet by SM textbooks? (See the discussion in subsections 10.5.1 on p. 129 and subsection 10.5.2 on p. 130.)

3-J.3 In physics, an explanation of an experiment by a theory does not deny the possibility that another theory will provide an entirely different explanation for the same event. The alternative explanation may be better. The three-jet event is an example of this issue. Two well-known cases are outlined below.

Let us examine the famous Michelson-Morley experiment, where a null result was obtained for the velocity of bodies through the ether. The assumption of length contraction of bodies and the time dilation of clocks explained the results of this experiment (see [129], pp. 3, 4). However, this explanation has been abandoned because the validity of SR is far beyond any doubt, and the ether concept has been removed.

The Bohr-Sommerfeld old quantum theory is another example of this issue. This theory provides a quite good explanation for the energy levels of the hydrogen-like atoms (see [56], pp. 34-42). However, this atomic theory cannot explain the states of atoms having more than one electron. Hence, the success of the Bohr-Sommerfeld atomic theory with the hydrogen-like atoms is meaningless, and this theory has been abandoned by the entire physical community (see e.g. the Dirac's lecture in [130]).

By the same token, one success of QCD does not prove its validity. In particular, this chapter presents many independent proofs of QCD failure. These proofs negate the three-jet argument by QCD supporters.

These points explain why one cannot deny the possibility that a contradiction-free theory will explain the three-jet event. As in many other cases discussed in this book, the RCMT does explain this event.

Conclusion: The large number of QCD failures under-
mines the QCD explanation of the three-jet events as
proof of the physical meaning of gluons. This effect
is another striking piece of evidence that electromagnetic fields are remarkably similar to the strong fields,
and the bremsstrahlung effect also appears in strong
interactions and not only in electromagnetic interactions. The RCMT is a monopole theory. As such,
it is a gluon-free theory claiming that strong forces
are similar to electromagnetic forces. This theory explains this three-jet event.

10.5.26 Remarks on QCD

This chapter has compared two strong-interaction theories –
RCMT and QCD. Using qualitative arguments, such as the existence or nonexistence of a state that a theory says that should
exist; The existence or non-existence of effects that a theory pre-
dicts; the increase or decrease of the cross-section of a scattering
experiment with the increase of the collision energy; the smaller or
larger volume of quarks and antiquarks in hadrons; and the smaller
or larger binding energy of particles. All examples have a solid experimental basis. Around 20 different effects support the RCMT
and refute QCD.

Readers are invited to consider this evidence and
make up their minds.

Chapter 11

Weak Interactions

This chapter analyzes the experimental and theoretical properties of weak interactions.

11.1 Experimental Properties of Weak Interactions

At low energy, weak interactions are detected in processes that involve neutrinos or violate flavor conservation and/or parity conservation. These processes are forbidden for strong and electromagnetic interactions. Several examples can be given of low-energy processes that are determined by weak interactions; for instance, the nuclear β decay, the $\mu^\pm$ decay, and the $\pi^\pm$ decay involve neutrinos, whereas some of the decay channels of the $K^\pm$ mesons do not contain neutrinos. Here, flavor conservation is violated and a strange quark disappears. More than 99.9% of the K_S^0 meson decays involve pions, and a strange quark disappears. More than 99% of the decays of the $\Lambda(1116)$ baryon comprise a nucleon and a pion. These are neutrinoless flavor-changing processes where a strange quark disappears.

Another clear weak interaction property is the *increase* of its intensity with the increase of energy. For example, Fig. 12.11 on p. 321 of [79] shows that the neutrino-nucleon cross-section is proportional to the collision energy. In contrast, the Rutherford, Mott, and Rosenbluth formulas prove that the electromagnetic cross-section *decreases* with energy. Hence, one expects that weak interactions will become significant for high-energy processes.

The decay data of the 80 GeV $W^\pm$ supports the weak interaction property outlined above and yields semi-quantitative information about this effect. About one third of the decay events are a charged lepton and a neutrino. The neutrino proves that these events are weak interaction processes. Most of the other decay modes contain hadrons. Charge conservation proves that among the outgoing particles, there is a charged hadron, and no antiparticle of this flavor compensates its charge. It means that we have a flavor-violating process because a charged hadron cannot be a particle-antiparticle state of the same flavor. Hence, the $W^\pm$ particle decay is a weak interaction process. The width of the $W^\pm$ is about 2.1 GeV. This quite large width means that the decay takes a short time. Furthermore, the different phase space between the two-lepton decay of the $W^\pm$ and that of the two-particle decay of the $\Lambda(1116)$ indicates that the intensity of weak interactions increases with energy. These data prove that at energy that is higher than 80 GeV, weak interactions are powerful.

The data of the $e^+e^- \to$ hadrons cross-section near the mass of the Z particle supports the previous conclusion (see e.g. [14], p. 711). The data show that at an energy of 87 GeV, the cross-section is 2.5 nb. The energy of 95.4 GeV is the corresponding point on the other side of the peak of the Z particle. Here, the cross-section is about 5.5 nb. This means that the Z-independent non-resonant e^+e^- cross-section *increases with the increase of energy,* and it reaches more than twice the corresponding value, at 87 GeV. The interaction of the e^+e^- collision begins as a combination of electromagnetic and weak interactions. It is well known that electromagnetic interaction *decreases* with the increase of energy. Hence, weak interactions are significant at energies of around 90 GeV. An analogous conclusion is inferred from Fig. 16.2 on p. 430 of [79].

11.2   The State of the Problem

The electroweak theory is the SM sector of weak interaction. As of today (year 2021), this theory is about 50 years old. Let us examine how fundamental principles of physics pertain to this theory. Physics is an experimental science, and the validity of any physical theory is supported by the fit of its predictions to data that belong to its domain of validity.

An experimental datum is verified if the state of a measurement device changes with time. This change of the measurement device reflects the time evolution of the system that is observed by the experiment. A physical theory describes the time evolution of a given system by means of time-dependent differential equations. Solutions of these equations describe specific cases. A few important examples of the significance of this theoretical structure are briefly presented in the following:

- Maxwell equations: Solutions to these equations have predicted electromagnetic radiation.

- The Lorentz law of force: Electric engines are constructed based on this law.

- The Einstein equations of general relativity: Solutions to these equations have predicted the bending of light rays passing near a massive body and the Schwarzschild radius, which is related to black holes.

- The Dirac equation: Solutions to this equation explain the electron's g-factor and the hydrogen atom energy levels, as well as predicting the existence of the positron.

Unlike the successful examples above, textbooks that discuss the electroweak theory do not show *an explicit form of its partial differential equations*. A fortiori, no solution to these equations has been reported, and no fit of these solutions to relevant data has been shown. Moreover, no SM textbook explains this grave omission.

> *Conclusion: About 50 years have elapsed since the construction of the electroweak theory but this theory still misses a vital element.*

This state of affairs calls for a thorough examination of experimental and theoretical foundations of weak interactions. This chapter is dedicated to this assignment. Readers may guess that the analysis proves more problematic aspects of the electroweak theory.

11.3 Landmarks of the Historical Progress of Weak Interactions

The following is a brief description of the historical progress of human knowledge about the weak interactions sector of physics; its observations yield instructive information:

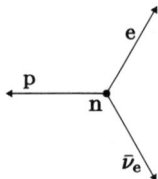

Figure 11.1: *Illustration of neutron β decay. Arrows denote the outgoing particles' momentum.*

WH.1 In 1930, Pauli postulated the existence of a spin-1/2 electrically neutral particle (later called neutrino) that is involved in the β decay.

WH.2 In 1934, Fermi proposed a weak interactions theory of the β decay of nuclei (see e.g. [79], p. 296). Figure 11.1 illustrates the neutron's β decay (see [23], p. 209). This figure shows that four fermions are involved in this weak process. The Fermi theory is based on a 4-fermion contact interaction, where a constant G_F with a dimension of $[L^2]$ describes the interaction strength. The Fermi β decay theory allows no change of the spin-projection s_z in the transition from the initial nuclear state to the final nuclear state.

WH.3 Gamow and Teller extended Fermi's work a few years later; their theory includes a β transition that allows a unity change of the nuclear angular momentum J. Experiments have shown that the strength of the Gamow-Teller transition is somewhat stronger than that of the Fermi transition:

$$G_A \simeq 1.3\,G_V, \tag{11.1}$$

where G_V, G_A denote the strength of the Fermi and the Gamow-Teller transitions, respectively (see [8], p. 188).

WH.4 Remark: The 4-fermion contact interaction of items WH.2 and WH.3 does not use a mediating field. In contrast, it is shown above that elementary classical and quantum particles are point-like (see subsection 4.3.1 on p. 46). Hence, an interaction between elementary particles requires a mediating field. For this reason, the Fermi and Gamow-Teller approach cannot be regarded as the last word.

WH.5 In 1956, Cowan, Reines, Harrison, Kruse, and McGuire reported the experimental detection of the neutrino.

WH.6 At about the same time, Lee and Yang proposed the idea that weak interactions violate parity conservation. In 1957, soon after the publication of this idea, Wu et al. carried out an experiment that proved parity violation in the β decay of the ^{60}Co nucleus (see [23], p. 214).

WH.7 The Wu experiment motivated Sudarshan and Marshak to examine the available weak decay data. They concluded that the weak interaction can be described as the difference between a vector and axial vector interaction (V-A) (see [23], pp. 217-220).

WH.8 In 1958, Feynman and Gell-Mann published an article that presented theoretical progress related to the problem of weak interactions [131]. Their work relied on the V-A finding of Sudarshan and Marshak and the idea of a massless neutrino. They also introduced the $(1 \pm \gamma^5)$ factor into the equations of motion. Here, the pure number 1 is a scalar, and the γ^5 matrix is a pseudoscalar (see [2], p. 26). Hence, $(1 \pm \gamma^5)$ is a maximal parity violation operator. This operator complies with the Sudarshan and Marshak V-A attribute of weak interactions.

WH.9 During the next decade, Glashow, Salam, and Weinberg formulated the electroweak theory, which is a field theory that combines electrodynamics and weak interactions. This theory uses the factor $(1 \pm \gamma^5)$, and the fields of the $W^\pm$, Z massive bosons mediate the weak interactions (see e.g. [79], chapter 15). According to this theory, the description of the Z boson is closely related to the electromagnetic 4-potential A_μ.

Let us examine a hypothetical historical course of weak interactions that is a variation of the real course described above. The alternative course departs from the real course at point WH.8. It relies on some of the experimental and theoretical elements that were already known in 1958:

New.8 The English translation of Wentzel's QFT textbook [132] was published in 1949. It explains how one can prove that conservation laws are satisfied by a QFT theory based on the variational principle. This approach is now regarded as the standard route for the construction of the QFT of a given particle. Indeed, it is stated that "all field theories used in current theories of elementary particles have Lagrangians of this form" (see [20], p. 300). Section 3.3 above discusses this topic.

New.9   The β decay is a time-dependent process. Hence, its theory should be based on an appropriate Hamiltonian.

New.10   The LHCD concept of section 7.4 says that a Lagrangian density is a Lorentz scalar, whereas a Hamiltonian density is the T^{00} entry of the energy-momentum tensor. Therefore, a term of the Lagrangian density and a corresponding term of the Hamiltonian density take different mathematical structures. This is a straightforward and indisputable assertion. Let us assume that the LHCD was already recognized in the 1950s.

New.11   Figure 11.1 shows that an electron and a neutrino are produced in a β decay. In 1958, knowledge about the electron was much better than that related to the neutrino. Hence, one naturally favors a weak interaction theory that relies on an appropriate extension of the electron's electromagnetic Lagrangian density (5.24). This means that the main weak interaction problem should be described as delineated below.

> *What is the form of the weak interaction term that should be added to the electron's electromagnetic Lagrangian density, where the corresponding Hamiltonian abides by the experimentally known properties of weak interactions?*

The required properties of such a weak interaction term are as follows:

Req_W.A   Following the Fermi constant G_F of weak interaction, the interaction strength should have the dimension $[L^2]$.

Req_W.B   Fermi describes weak interactions by means of four fermions and no mediating field. The point-like attribute of elementary particles and the relative success of the Fermi theory mean that weak interactions should have a short range.

Req_W.C   Following item New.10, the associated Hamiltonian should comply with the *V-A* property of weak interactions.

New.12   In 1941, Pauli examined the following interaction term of the electron (see [52], p. 223):

$$\mathcal{L}' = \mathrm{d}\bar{\psi}\sigma_{\mu\nu}F^{\mu\nu}\psi, \tag{11.2}$$

where $F^{\mu\nu}$ is the electromagnetic field tensor, the coefficient d has the dimension of length, and

$$\sigma_{\mu\nu} \equiv \frac{i}{2}(\gamma^\mu\gamma^\nu - \gamma^\nu\gamma^\mu). \qquad (11.3)$$

This term is called tensor interaction because of its dependence on $\sigma_{\mu\nu}$. The interaction (11.2) alters the Dirac expression for the electron's dipole moment (see [20], p. 14; [52], p. 223). However, the ordinary Dirac Lagrangian density (5.17), which contains no term like (11.2), yields a good prediction for the electron's magnetic dipole moment. Hence, the Pauli term (11.2) has been abandoned as a term that pertains to the electron's *electromagnetic interaction*.

Favorable words about the Pauli term can be found in the present literature. For example, Weinberg asserts that the Pauli term "is consistent with all accepted invariance principles, including Lorentz invariance and gauge invariance, and so there is no reason why such a term should not be included in the field equations" (see [20], p. 14).

The author of this book proved that the Pauli term (11.2) is the required weak interaction term of the electron's Lagrangian density (see [133–135]). Here, the external field $\mathcal{F}^{\mu\nu}$ is a Maxwellian-like weak field of a weak dipole that is associated with the spin of an external Dirac particle. The coefficient d of the Pauli term has the dimension $[L]$. Therefore, this two-body interaction strength has the dimension $[L^2]$. Moreover, a proof showing that the associated Hamiltonian abides by the *V-A* property of weak interactions has been published [133–135]). The Lagrangian density of the alternative theory is

$$\mathcal{L}_D = \bar{\psi}(\gamma^\mu i\partial_\mu - m)\psi - e\bar{\psi}\gamma^\mu A_\mu\psi + \mathrm{d}\bar{\psi}\sigma_{\mu\nu}\mathcal{F}^{\mu\nu}\psi, \qquad (11.4)$$

where the last term is the Pauli term (11.2). The Lagrangian density (11.4) is an extension of the Dirac QED Lagrangian density (5.24) and the last term of (11.4) represents weak interactions. Below, this theory is called *Dipole-Dipole Weak Interaction Theory* (DDWIT).

New.13   The real historical course of the weak interaction introduced the $(1 \pm \gamma^5)$ factor to *impose* the *V-A* attribute. In contrast, the alternative route described here uses general principles of physics, where the Hamiltonian that is derived from the

Lagrangian density (11.4) *proves* the *V-A* weak interactions attribute. Certainly, this is a better theoretical structure.

> Conclusion: The alternative historical progress that is described in items New.8–New.13 relies on elements that were nearly all already known in 1958. The missing element is the relativistic difference between the Lorentz scalar attribute of the Lagrangian density and the Hamiltonian density of item New.10, which is the T^{00} component of the energy-momentum tensor.

Indeed, the quite straightforward proof of the validity of the theoretical element 7.4 that is called LHCD here has apparently been unnoticed until the author of this book published it in [133]. This evidence looks like the reason for the rejection of the tensor interaction (11.2) as the basis for weak interactions (see the discussion near (8) of [131] or [23], p. 219). This fundamental error is the reason for the construction of the electroweak theory.

Two problems stem from this situation:

- Does the electroweak theory have a coherent structure?

- Is the Pauli term of (11.4) a coherent basis of weak interactions?

The next sections of this book discuss these problems.

11.4   The Weak Interaction Problem

The data in Tables 9.1 and 9.2 can be used not only for the unification of strong and electromagnetic interactions but also for treating weak interactions by means of a different mathematical structure. For the convenience of the discussion, let us copy these tables here.

Table 11.1: Interaction Unification in the RCMT

Theory	Interaction	Parity Conserv.	Flavor Conserv.	Photon Interaction
RCMT	Strong	YES	YES	YES
	EM	YES	YES	YES
DDWIT	Weak	NO	NO	NO

Table 11.2: Interaction Unification in the SM

Theory	Interaction	Parity Conserv.	Flavor Conserv.	Photon Interaction
QCD	Strong	YES	YES	YES
Electroweak	EM	YES	YES	YES
	Weak	NO	NO	NO

Consideration of these tables confirms that well-established experimental data do not recommend a unification of the electromagnetic and weak interactions. Relying on this experimental information, the DDWIT uses the Lagrangian density (11.4) (see p. 177). The weak interaction term of this Lagrangian density is a dipole-dipole interaction. Hence, its mathematical form is not the same as that of the QED electromagnetic term or the RCMT strong interaction term (10.3) (see p. 125). These theories use charges and monopoles – namely, Lorentz scalars – whereas the DDWIT uses a spin-related dipole of an elementary point-like Dirac particle.

Alternatively, the electroweak theory unifies the electromagnetic and weak interactions. This course clearly goes against the data shown in Table 11.2. Below, several sections of this chapter prove erroneous electroweak elements. In so doing, they support a general dictum stating that ignoring well-established experimental data is not a good scientific practice.

11.5 The Dipole-Dipole Weak Interaction Theory

A coherent weak interaction theory should abide by relevant well-established experimental data. Here, let us rephrase the points of New.11, on p. 176:

WD.1 The Fermi constant, G_F, that describes the weak interaction strength, has the dimension $[L^2]$.

WD.2 The Fermi description of weak interactions uses four fermions and no mediating field. The point-like attribute of elementary particles and the relative success of the Fermi theory mean that weak interactions should have a short range.

WD.3 The β decay has the Sudarshan and Marshak parity violation feature, which is denoted by V-A (vector minus axial vector).

Let us see whether the Pauli term (11.2) of item New.12 satisfies these requirements. For convenience, this term is rewritten here as

$$\mathcal{L}' = d\bar{\psi}\sigma_{\mu\nu}\mathcal{F}^{\mu\nu}\psi, \tag{11.2}$$

This interaction term says that weak interaction is an interaction between dipoles of Dirac particles. Here, the constant d of (11.2) denotes the strength of a dipole, and its dimension is $[L]$. In addition, the strength of the field $\mathcal{F}^{\mu\nu}$ of another particle is proportional to d. Hence, the strength of the Pauli term (11.2) takes the required dimension $[L^2]$.

The fields tensor $\mathcal{F}^{\mu\nu}$ of (11.2) emanates from a weak dipole, and its mathematical expressions can be readily taken from electrodynamics. Let us write down the required expressions. The electron's spin is an axial dipole. In electrodynamics, the field of a motionless point-like axial dipole $\boldsymbol{\mu}$ is

$$\boldsymbol{B} = [3(\boldsymbol{\mu} \cdot \boldsymbol{r})\boldsymbol{r} - r^2\boldsymbol{\mu}]/r^5. \tag{11.5}$$

Hence, the tensorial form of this field is

$$F^{\mu\nu} = \begin{pmatrix} 0 & 0 & 0 & 0 \\ 0 & 0 & -B_z & B_y \\ 0 & B_z & 0 & -B_x \\ 0 & -B_y & B_x & 0 \end{pmatrix}. \tag{11.6}$$

The neutrino mass is extremely small. This means that a neutrino is an ultrarelativistic particle. Hence, the Lorentz transformation that describes an ultrarelativistic particle that moves in the x-direction is

$$\Lambda^\mu_\nu = \begin{pmatrix} \eta & \theta & 0 & 0 \\ \theta & \eta & 0 & 0 \\ 0 & 0 & 1 & 0 \\ 0 & 0 & 0 & 1 \end{pmatrix}. \tag{11.7}$$

Here $\eta \gg 1$, and $\theta^2 = \eta^2 - 1$. Hence, $\theta \simeq \eta \gg 1$.

Applying the Lorentz transformation (11.7) to the fields tensor (11.6), one finds the neutrino's electromagnetic-like fields tensor

$$F^{\mu\nu}_{LT} = \begin{pmatrix} 0 & 0 & -\theta B_z & \theta B_y \\ 0 & 0 & -\eta B_z & \eta B_y \\ \theta B_z & \eta B_z & 0 & -B_x \\ -\theta B_y & -\eta B_y & B_x & 0 \end{pmatrix}. \tag{11.8}$$

In other words, field components that are perpendicular to the boost direction become very large, and the electric and magnetic fields take about the same value. The field component that is parallel to the boost direction can be ignored (see [3], pp. 98-100). These conclusions concerning the electromagnetic fields also apply to the Maxwellian-like fields of the weak dipole of a Dirac particle.

The β decay is a time-dependent process. Hence, the Hamiltonian's form of the Pauli term (11.2) is required. For this end, let us examine the identity

$$\bar{\psi} \equiv \psi^\dagger \gamma^0 \tag{11.9}$$

(see [2], p. 24). In the Dirac theory, $\psi^\dagger \psi$ is the particle's density. Moreover, the Dirac ψ is the generalized coordinate of the Lagrangian density, whereas $i\psi^\dagger$ is its conjugate momentum (see [14], p. 52). Hence, the Hamiltonian density should be written in terms of ψ, $\psi^\dagger$. The calculation below proves that γ^0 of (11.9) changes dramatically the form of the Pauli term (11.2). Substituting (11.9) into (11.2) and using $\gamma^0 = \gamma_0$, one finds

$$
\begin{aligned}
d\psi^\dagger \gamma^0 \sigma_{\mu\nu} \mathcal{F}^{\mu\nu} \psi &= 2id\psi^\dagger (\gamma_0 \gamma_0 \gamma_i \mathcal{E}^i + \gamma_0 \gamma_1 \gamma_2 \mathcal{B}^3 - \gamma_0 \gamma_1 \gamma_3 \mathcal{B}^2 \\
&\quad + \gamma_0 \gamma_2 \gamma_3 \mathcal{B}^1)\psi \\
&= 2id\psi^\dagger (\gamma_i \mathcal{E}^i - \gamma_0 \gamma_1 \gamma_2 \gamma_3 \gamma_3 \mathcal{B}^3 + \gamma_0 \gamma_1 \gamma_3 \gamma_2 \gamma_2 \mathcal{B}^2 \\
&\quad - \gamma_0 \gamma_2 \gamma_3 \gamma_1 \gamma_1 \mathcal{B}^1)\psi \\
&= 2d\psi^\dagger (i\gamma_i \mathcal{E}^i - \gamma^5 \gamma_3 \mathcal{B}^3 - \gamma^5 \gamma_2 \mathcal{B}^2 - \gamma^5 \gamma_1 \mathcal{B}^1)\psi \\
&= 2d\psi^\dagger (i\gamma_i \mathcal{E}^i - \gamma^5 \gamma_i \mathcal{B}^i)\psi \tag{11.10}
\end{aligned}
$$

In the second line of (11.10), three terms are multiplied by $1 = -\gamma^k \gamma^k$, $k \neq i$, $k \neq j$. In the third line, the γs are reordered and the anti-commutation relation $\gamma^\mu \gamma^\nu + \gamma^\nu \gamma^\mu = 2g^{\mu\nu}$ is used. The pseudoscalar $\gamma^5 \equiv i\gamma^0 \gamma^1 \gamma^2 \gamma^3$ is substituted.

This argument explains why the result of (11.10) makes sense. It ignores the sign of the final result. The tensor $\sigma_{\mu\nu}$ comprises six terms – three of which take the form $\gamma^0 \gamma^i$ and the other three are $\gamma^i \gamma^j$, where $i \neq j$. The relation $(\gamma^0)^2 = 1$ proves that multiplication by γ^0 casts a term of the first kind into γ^i. On the other hand, multiplication of a term of the second kind by γ^0 yields $\gamma^0 \gamma^i \gamma^j$. The multiplication of the later product by $(\gamma^k)^2 = -1$, $(k \neq i, k \neq j)$ yields an axial vector whose form is $\gamma^5 \gamma^k$.

The three γ^i matrices of the first term of the last line of (11.10) are antihermitian. The imaginary factor i means that the first term of this line is Hermitian. This weak interaction term is analogous to the electromagnetic term of the three Dirac α^i matrices, which are used in the Dirac electromagnetic Hamiltonian. The following calculation proves that the product of the γ matrices of the second term of the last line of (11.10) is also Hermitian:

$$(\gamma^5\gamma^i)^\dagger = \gamma^{i\dagger}\gamma^{5\dagger} = -\gamma^i\gamma^5 = \gamma^5\gamma^i. \qquad (11.11)$$

Hence, the two terms of (11.10) are Hermitian operators that correspond to the vector $\boldsymbol{V}$ and the axial vector $\boldsymbol{A}$ parts of the weak interactions. These Hermitian operators are suitable for the Hamiltonian. The very small mass of the neutrino means that it moves ultra-relativistically, and $|\mathcal{E}| \simeq |\mathcal{B}|$. The equality of the neutrino's electric-like and magnetic-like fields of (11.8) means that these terms are contracted with 3-vectors that practically have the same absolute value. This means that the weak interaction theory derived above *proves* that weak processes do not conserve parity. This result is also consistent with the equal weight of $\boldsymbol{V}$ and $\boldsymbol{A}$ in the well-known *V-A* form of leptonic weak interactions (see [23], p. 220), meaning that the dipole structure of the weak interaction theory developed herein *proves* that an interaction of a neutrino with a Dirac particle is in accordance with the parity-violating *V-A* form of weak interactions.

The history of the Pauli term can briefly be described as follows. Pauli has formulated this term as an additional term of the electromagnetic interaction of the electron [52]. It has been abandoned because it is unsuitable for this purpose. Taking an abstract point of view, Weinberg examined this term, stating, "A more modern approach would be simply to remark that the term (1 .1 .32) is consistent with all accepted invariance principles, including Lorentz invariance and gauge invariance, and so there is no reason why such a term should *not* be included in the field equations" (see [20], p. 14). (Please note that in [20], (1.1.32) denotes the Pauli term.) Years later, this term was rediscovered by the author of this book as an element of a consistent weak interaction theory [133].

The weak interaction term is a part of every spin-1/2 Dirac particle, including quarks, charged leptons, and neutrinos. Thus, the QED Lagrangian density (3.33) of a charged lepton should be upgraded to include the weak interaction Pauli term (11.2). Its explicit form is

$$\mathcal{L}_\ell = \bar{\psi}[\gamma^\mu i\partial_\mu - m - e\gamma^\mu A_\mu - \mathrm{d}\sigma_{\mu\nu}\mathcal{F}^{\mu\nu}]\psi. \qquad (11.12)$$

In other words, the first term of (11.12) represents dynamical properties, the second term is the mass term, the third term represents electromagnetic interactions, and the last term represents weak interactions. Note the analogy between the interaction terms of (11.12): Each of them is a product of three factors – a coefficient that denotes the intensity of the interaction and a tensorial contraction of an expression of Dirac γ matrices with an external field.

The weak interaction dipole field (11.5) decreases like r^{-3}. Remembering that the electromagnetic interaction term depends on the potential that decreases like r^{-1}, one can conclude that the DDWIT proves that the range of weak interaction is much shorter than that of the electromagnetic interaction. This feature is compatible with requirement WD.2, given at the beginning of this section.

The weak interaction dipole field *differs* from the electromagnetic dipole fields. The fundamental electromagnetic term is the particle's charge. Here, electric and magnetic dipoles arise from a power series expansion of the fields (see e.g., [28], pp. 391-401). The calculation proves that electromagnetic radiation fields are associated with these dipoles. In contrast, the fields of the weak dipoles decrease like r^{-3}.

Conclusion: Unlike electromagnetic fields, weak interaction fields have no radiation component.

This conclusion is compatible with the short-range attribute of weak interactions.

11.6 Problems with the Electroweak Theory

The historical progress that has culminated with the construction of the electroweak theory proves that this theory stems from a mathematically erroneous concept that ignores the intrinsic tensorial differences between the Lagrangian density and the Hamiltonian density. Moreover, Table 11.2 proves that the electroweak theory goes against well-established data. Fundamental principles of this book state that a physical theory should have a coherent mathematical structure (see Section 3.1 on p. 12) *and* be compatible with well-established data that belong to its domain of validity (see item Req.3 on p. 2). This state of affairs motivates a critical and thorough examination of the electroweak theory. This section is dedicated to this pursuit, beginning with a presentation of a list of erroneous elements of the electroweak theory.

11.6.1   The Electroweak Equation of Motion

Section 11.2 mentioned the lack of an explicit equation of motion
for the quantum particles of the electroweak theory. This issue
was used as a reason for a judicious weak interactions analysis.
However, the lack of this vital element is certainly an erroneous element of the electroweak theory. Hence, it is also on this list. Other
problems of the electroweak $W^\pm$ and the Z bosons are mentioned
below.

11.6.2   Problems with the Electroweak's De-scription of the $W^\pm$ Bosons

The $W^\pm$ particles are vital elements of the electroweak theory. This
theory regards the $W^\pm$ function as a 4-vector that is analogous to
the electromagnetic 4-potential A_μ. Let us examine a term of the
W^+ Lagrangian density of the electroweak theory:

$$\mathcal{L}_W = -\frac{1}{2}|\bar{D}_\mu W_\nu^+ - \bar{D}_\nu W_\mu^+|^2 + ..., \qquad (11.13)$$

where D_μ is

$$\bar{D}_\nu = \partial_\mu + O_\mu. \qquad (11.14)$$

Here O_μ is a 4-vector that is irrelevant to the present discussion (see
[47], p. 518). The Lagrangian density (11.13) contains a term that
is a product of two derivatives of the W_μ function. This property
yields the two following results:

- As shown in section 3.3, the dimension of the Lagrangian density is $[L^{-4}]$. Hence, the dimension of the quantum function
  W_μ is $[L^{-1}]$. In contrast, it is proved in chapter 3.2 that the
  dimension of the Schroedinger function ψ is $[L^{-3/2}]$. Therefore, the electroweak description of the W_μ function violates
  the correspondence principle of chapter 3.2.

- Section 3.4 shows how the Noether theorem yields an expression for a conserved 4-current of a quantum particle (3.14).
  For convenience, let us rewrite it here:

$$j^\mu = \frac{\partial \mathcal{L}}{\partial \psi_{,\mu}} \psi. \qquad (3.14)$$

  The discussion of section 3.4 proves that the 4-current obtained for the $W^\pm$ contains derivatives. Maxwellian electrodynamics proves that a quantum function of a charged particle, like $W^\pm$, requires a coherent derivative-free expression
  for the 4-current.

Mathematical constraints are stern and unforgivable. The fate of the electroweak description of the $W^\pm$ electromagnetic interaction provides a good illustration of this issue. Let us begin with the Dirac theory of the electron. This is a good theory of an elementary charged particle. Thus, about one month after the publication of the Dirac theory of the electron [44], Darwin found an expression for the conserved 4-current of this particle [136].

The historical fate of the electrically charged $W^\pm$ is completely different. The electroweak theory was formulated before 1970. About two decades elapsed before a group of researchers published papers that proposed to treat the electromagnetic interaction of $W_\mu^\pm$ by means of an *effective Lagrangian density* [137, 138]. Their effective formulas are still used by people who work at large research centers like Fermilab and CERN [139, 140]. The expression for the electromagnetic interaction of these articles depends on a derivative of the quantum function $W_\mu^\pm$ (see Eq. (3) of [140]). Hence, it is unacceptable as a coherent description of electromagnetic interaction (see section 6.2). This means that a coherent description of the electroweak $W^\pm$ 4-current is still unavailable.

Remark: it is interesting to note that besides the fundamental error with $W^\pm$ mentioned above, the effective Lagrangian density of the Fermilab and CERN publications [139, 140] contains a term that violates a well-known principle of physics called the balance of dimension of all terms of a physical expression. Thus, the electromagnetic term that stands on the right-hand side of the second line of Eq. (3) of [140] can be put in standard electromagnetic notation as

$$-ieW_\mu^\dagger W_\nu F^{\mu\nu}. \qquad \text{(A term of Eq. (3) of [140])}$$

The Maxwellian electromagnetic field $F^{\mu\nu}$ of this expression is charge-free, and the dimension of the $W_\mu^\pm$ is $[L^{-1}]$. However, an electromagnetic interaction term of a Lagrangian density should be proportional to *charge density*. Hence, the overall dimension of its *charge-carrying functions* must be $[L^{-3}]$. In contrast, the dimension of the product $W_\mu^\dagger W_\nu$ is $[L^{-2}]$. This means that this term in [140] represents an electromagnetic interaction where the strength is not proportional to the particle's charge density! Unfortunately, the same erroneous term is also found in [137–139]. The number of authors of [137–140] is very large, and it takes four decimal digits. Apparently, none of these persons has detected this gross error.

> Conclusions: In the case of a correct theory of a
> charged particle, such as the Dirac theory of the elec-
> tron, researchers found a consistent expression for the
> particle's 4-current right away. In contrast, even af-
> ter several decades, SM supporters still use a theoret-
> ically inconsistent expression for a description of the
> electromagnetic interaction of the electroweak's $W^{\pm}$
> particles. This means that, unlike the Dirac theory
> of the electron, the electroweak theory is inherently
> wrong.

11.6.3 More Problems with the Electroweak's Description of the $W^{\pm}$ Bosons

Before entering into detail, it is important to emphasize a crucial
theoretical element of electrodynamics. The electromagnetic Lagrangian density is the cornerstone of the theory that describes
the time evolution of *two* kinds of physical objects – the electromagnetic fields *and* the charged particles. The celebrated textbook [3] discusses the classical theory in detail and proves this
property: Maxwell equations are derived for the electromagnetic
fields (see [3], pp. 70,71,78,79), and the Lorentz force is proved
for the charged particle (see [3], p. 51). This issue is also shown
in QED, where one uses the Lagrangian density (3.33) (see p. 35
of this book). This Lagrangian density is presented here for the
convenience of the discussion:

$$\mathcal{L}_{QED} = \bar{\psi}[\gamma^{\mu}i\partial_{\mu} - m]\psi - eA_{\mu}\bar{\psi}\gamma^{\mu}\psi - \frac{1}{16\pi}F_{\mu\nu}F^{\mu\nu}. \qquad (11.15)$$

The changes of the quantum version of the Lagrangian/Lagrangian
density are a replacement of the classical self-term of a particle by
the self-terms of a Dirac particle that are enclosed in the square
brackets of (11.15). The interaction term also changes. Here, the
4-current of the Dirac particle,

$$j^{\mu} = \bar{\psi}\gamma^{\mu}\psi \qquad (11.16)$$

replaces the classical 4-current.

Both CPH and QED prove that the laws of the time evolution
of the charged particle *and* the electromagnetic fields are derived
from *the same interaction term*. QED uses the penultimate term
of (11.15), and CPH uses its classical analog (see [3], pp. 48, 75).

This is an *extremely* important attribute of the electromagnetic theory because it proves the theoretical coherence of electrodynamics. Therefore, it deserves a specific name: *The Dual Attribute of Electromagnetic Interactions* (DAEI). A relevant point is described in item CNST.9 on p. 41.

Remembering DAEI, let us examine other aspects of the electroweak description of the $W^{\pm}$ particles. In addition to the problems that were mentioned in the previous subsection, the term (11.13) of the $W^{\pm}$ electroweak Lagrangian density suffers from other kinds of contradictions. It has a quadratic expression of the electromagnetic 4-potential that takes the following form:

$$\mathcal{L}_{EW} = \eta e^2 A^\mu A_\mu W^{+\nu} W_\nu^+ + ..., \qquad (11.17)$$

where η denotes a numerical constant. Furthermore, factor e^2 proves that this interaction term is not proportional to electric charge e. This term is obtained from an expansion of the first term on the second line of Eq. (C.18) on p. 518 of [47]. An analogous expression is shown in Eq. (11.31) on p. 113 of [48].

The electromagnetic 4-potential A_μ is regarded as the generalized coordinate of the electromagnetic fields Lagrangian density (see [3], p. 78). It is well known that the Euler-Lagrange equation (3.9) on p. 21 comprises a derivative in terms of the generalized coordinate. Hence, the term (11.17) proves that the electroweak version of the equation of motion of the electromagnetic fields is

$$F^{\mu\nu}_{,\nu} = \eta' e^2 A^\mu W^{+\lambda} W_\lambda^+ + ..., \qquad (11.18)$$

where η' denotes another numerical constant. Result (11.18) comprises *two* gross violations of Maxwellian electrodynamics. The right side of the correct version of the inhomogeneous Maxwell equations of the electromagnetic fields (3.24) is proportional to electric charge e *and* it is independent of the 4-potential A_μ (see [3], p. 79; [20], pp. 341-342). In contrast, (11.18) is the electroweak version of the equations of the electromagnetic fields. It violates the Maxwell equations because its right side contains factor e^2, and it *explicitly* depends on the electromagnetic 4-potential. The second point means that the electroweak theory violates gauge invariance.

Conclusion: The electroweak theory of the electrically charged $W^{\pm}$ particles is inherently wrong because it violates Maxwellian electrodynamics in general and gauge invariance in particular.

11.6.4   On the Significance of the DAEI

This subsection shows several proofs that the present particle
physics community simply does not understand the profound meaning of the DAEI. Before doing this, let us try to see the meaning
of this discrepancy.

Physics has already acquired a substantial number of principles
that are regarded as vital elements of any relevant physical theory.
It is hard to state which principle is more meaningful, and this
issue is probably not a good question. However, one may introduce a minor change into the form of this problem and ask which
principle is more effective for *physical research*. In this case, a rule
of thumb says that an unknown principle is much more important
than a known principle is. This opinion relies on the expectation
that people will abide by a well-known principle. In contrast, an
unknown principle is doomed to be ignored, and detrimental results
can follow.

The KG equation was reintroduced in the 1930s by Pauli and
Weisskopf [141]. Their Lagrangian density has a term that is analogous to the troublesome term of the electroweak's $W^\pm$ (11.17) –
namely, it has a factor e^2 and a product of the electromagnetic 4-
potential $A_\mu A^\mu$ (see Eq. (37) on p. 198 of [141]). This means that
the proof of the previous subsection applies to the KG equation.

> Conclusion: The KG theory of an electrically charged
> particle is inherently wrong because it violates
> Maxwellian electrodynamics in general and gauge in-
> variance in particular.

Contrary to this conclusion, nearly every mainstream QFT textbook presents the KG equation as a coherent physical theory, and
its inherent errors are not mentioned! This state of affairs proves
that the profound meaning of the DAEI requirement is still unnoticed by mainstream particle physicists.

Referring to the electroweak $W^\pm$ problem of the previous section, it can be stated that mainstream textbooks support the electroweak theory, but most of them take a strange approach and
refrain from showing the $W^\pm$ electromagnetic interaction term of
the theory's Lagrangian density. A fortiori, they do not discuss the
consequences of the DAEI requirement. This requirement, which
was presented in the previous section, refutes the electroweak theory.

Recently, I have exchanged several emails with a distinguished
particle physics expert, whom I refer to as X here. Section 16.2

contains details of this correspondence. It turns out that also X is unaware of the DAEI.

> Conclusion: The evidence described above substantiates the assertion that the present particle physics community simply does not understand the profound meaning of the DAEI.

11.6.5 Problems with the Electroweak's Z boson

The Z particle plays a crucial role in the electroweak theory. It is a massive particle, and its mass is about 91 GeV. Like the electromagnetic 4-potential, its electroweak wave function takes the form of a 4-vector where the entries are mathematically real functions. Please note that section 7.3 on p. 76 proves that a mathematically real quantum function cannot describe a massive particle. This is yet another inherent contradiction of the electroweak theory.

11.6.6 The Significance of Density

It is proved in subsection 11.6.2 that the electroweak theory cannot provide a coherent expression for the $W^\pm$ 4-current. Density is the 0-component of the 4-current. Hence, the electroweak theory has no expression for the $W^\pm$ density. The Z particle suffers from the same problem. The electroweak theory uses a mathematically real function for this particle. It is proved in section 4.2 that density cannot be defined for a mathematically real function of a quantum particle.

The significance of density can be inferred from the experiments depicted on Fig. 11.2. The outgoing particles are Dirac particles produced by the decay of the W^-, Z. In principle, these particles can be detected by devices that measure their space-time position and their energy-momentum at the instant of detection. The Dirac

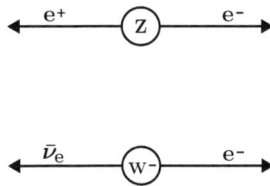

Figure 11.2: *Leptonic decay channels shown in the rest frame of the Z and the W^-.*

theory provides coherent expressions for the particle's 4-current
and density. The particle's energy and momentum are also defined.
The measured data of each of the outgoing particles show that the
two particles were emitted from the same small space-time region
and the combined energy of the outgoing particles is consistent
with the mass of the decaying particle. Therefore, experimenters
have concluded that the two outgoing particles are decay products
of the $W^{\pm}$, Z particles, respectively. This analysis depends on a
coherent definition of the density of each of the decaying particles
and its energy-momentum.

> Conclusions: The experimental definition of the
> $W^{\pm}$, Z particles depends on a good definition of their
> position and energy. A coherent expression for den-
> sity is the quantum expression of the particle's posi-
> tion. The absence of a coherent density expression
> for the electroweak theory of the $W^{\pm}$, Z means that
> this theory fails to describe crucial experiments where
> these particles are measured.

Another aspect of this issue is the canonical interrelations between position and momentum, which yield the quantum uncertainty relation. Evidently, to show position-momentum uncertainty
relations, one must have a quantum expression for density, where
the spatial integral yields an expression for the mean value of the
particle's position.

11.6.7   Consequences of the Massive Neutrino

It is now recognized that "neutrinos can no longer be considered
as massless particles" [142]. Their spin-1/2 means that they are
described by a 4-component Dirac-like spinor. (The neutrino is not
a Majorana particle because the Majorana neutrino has inherent
contradictions. See section 12.2 on p. 203.) Let us examine the
meaning of this outcome.

Pair production and pair annihilation are important properties
of a Dirac particle. Many experiments show these processes for the
electron and the muon. The recognition of the neutrino as a Dirac
particle means that neutrinos can be involved in these processes.

Factor $(1 \pm \gamma^5)$ is widely used in interaction terms of the electroweak theory (see e.g [23], pp. 219, 220; [14], p. 559; [46], p.
308). This expression shows how this factor operates on quantum
functions

$$< \bar{\psi}_\ell | \hat{O}(1 \pm \gamma^5) | \psi_\nu > \qquad (11.19)$$

Here $\hat{O}$ represents an appropriate operator, and the subscripts ℓ, ν denote a massive charged lepton and a neutrino, respectively. It turns out that this expression does not hold for a massive Dirac neutrino. Indeed, operating with $(1 \pm \gamma^5)$ on a motionless spin-up Dirac spinor, one obtains

$$
\begin{pmatrix}
1 & 0 & \pm 1 & 0 \\
0 & 1 & 0 & \pm 1 \\
\pm 1 & 0 & 1 & 0 \\
0 & \pm 1 & 0 & 1
\end{pmatrix}
\begin{pmatrix}
1 \\
0 \\
0 \\
0
\end{pmatrix}
=
\begin{pmatrix}
1 \\
0 \\
\pm 1 \\
0
\end{pmatrix}. \tag{11.20}
$$

Here the notation of the γ matrices is that of [2], pp. 17, 25.

The right side of (11.20) is a Dirac spinor that has an infinite energy-momentum (see [2], p. 30). This means that the operator $(1 \pm \gamma^5)$ casts a motionless Dirac particle into an unphysical state. Furthermore, a product of two γ matrices is used for a boost of a Dirac particle (see [2], p. 21). Hence, $(1 \pm \gamma^5)$ commutes with the boost operator. For this reason, the operator $(1 \pm \gamma^5)$ casts *any* Dirac spinor into the unphysical state of an infinite energy-momentum. Such a quantum state is not included in a Hilbert space. Hence, the required matrix elements cannot be calculated. Relying on the correspondence relationships between QFT and QM (see section 3.2), one can conclude that the problem that arises from the factor $(1 \pm \gamma^5)$ holds for QM and the QFT electroweak theory as well.

Factor $(1 \pm \gamma^5)$ is used for adapting the SM to the experimental $V - A$ property of weak interactions (see [23], pp. 217-220). The contradiction obtained above indicates that the SM is inconsistent with the $V - A$ property of weak interactions.

A further examination of the electroweak factor $(1 \pm \gamma^5)$ of (11.19) proves that the electroweak contradiction mentioned above is an old one. To see this, let us examine one term of the operator $\hat{O}$ of (11.19). This quantity comprises a product of n γ^μ matrices, where $0 \le n \le 4$. The unit matrix commutes with any matrix, and the matrix γ^5 commutes with the product of every even number of γ^μ matrices; it anticommutes with the product of every odd number of γ^μ matrices. Therefore, a change of the order of operator $\hat{O}$ and factor $(1 \pm \gamma^5)$ of (11.19) yields either the same expression

$$
< \bar{\psi}_\ell | (1 \pm \gamma^5) \hat{O} | \psi_\nu > \tag{11.21}
$$

or the modified expression

$$
< \bar{\psi}_\ell | (1 \mp \gamma^5) \hat{O} | \psi_\nu > . \tag{11.22}
$$

In quantum theories, every term of the Hamiltonian operator is Hermitian. For this reason, in (11.21) and (11.22), the operator $(1 \pm \gamma^5)$ operates on the function $\bar{\psi}_\ell$ of a charged massive lepton, which is the $< bra|$ of these expressions.

> Conclusion: Factor $(1 \pm \gamma^5)$ is a vital element of the electroweak theory. In the case of a massive neutrino, this factor yields a contradiction. The same contradiction holds for the electron and the other electrically charged massive leptons.

11.6.8   The Occam's Razor and the Complexity of the Electroweak Theory

It is agreed that the Lagrangian density is the cornerstone of QFT (see section 3.3). Let us compare the Lagrangian density of the DDWIT (11.12) with that of the electroweak theory. The DDWIT Lagrangian density comprises *four terms* that describe electromagnetic and weak interactions of a Dirac particle. (The addition of the self-term of the electromagnetic fields $-\frac{1}{16\pi}F^{\mu\nu}F_{\mu\nu}$ shows that the full Lagrangian density of these interactions has five terms.)

It turns out that many theoretical textbooks that discuss the electroweak theory *do not show an explicit form of the full Lagrangian density of this theory.* The textbook [47] is one exception. On p. 522, it shows the Lagrangian density of the electroweak's bosons. This textbook uses these words just before the Lagrangian density. "The complete result for the bosonic fields, including the vector density from Eq. (C.18) above, is of a still lengthy, but more standard, form." However, the Lagrangian density in [47] is incomplete because it does not show how spin-1/2 Dirac particles interact with its bosons. Moreover, the electroweak Lagrangian density of [47] comprises more than 30 terms! (The terms that compose its $\mathcal{L}_{vh}$ are counted.) Similarly, the section *After electroweak symmetry breaking* of the present (August 2021) Wikipedia entry on the electroweak theory shows the same kind of picture. Here, the electroweak Lagrangian density $\mathcal{L}_{EW}$ is written as a sum of eight sub-Lagrangians. Summing up the terms of these sub-Lagrangians, one finds far more than 30 terms! The textbook [48], pp. 113, 114 shows an analogous picture.

The extremely complicated electroweak Lagrangian density described above certainly yields even more complicated Euler-Lagrange equations. These are the equations of motion of this theory. Unlike the QED equations, no textbook contains an adequate discussion of the electroweak equations. A fortiori, no electroweak textbook shows the fit of solutions of these equations to experimental data. Moreover, electroweak textbooks do not explain why the theory's equations of motion are not discussed appropriately.

> Conclusion: The Occam's razor principle [143] favors simplicity. Hence, it casts serious doubt on the physical merits of the electroweak theory, where the Lagrangian density comprises a multitude of terms. Support for this conclusion can be inferred in that many electroweak textbooks take an unusual policy and refrain from writing down an explicit form of the electroweak's full Lagrangian density and its Euler-Lagrange equations. Furthermore, the remark mentioned earlier in this book is relevant to the present case: "If one theory relies on six assumptions and the other theory relies on more than thirty assumptions, then the Occam's razor criterion is decisive!"

11.7 A Comparison Between Weak Interaction Theories

Two different weak interaction theories are shown above – the electroweak theory and the DDWIT. This section shows the overwhelming advantage of the DDWIT over the electroweak theory. Arguments showing many cases where the electroweak theory is inconsistent with well-established physical laws have already been pointed out above. They are mentioned here to stress the difference between the two competing theories as follows:

1. The electroweak theory relies on the assumption of a massless neutrino. Few examples that are taken from the literature support this claim: Salam said in his Nobel lecture that the electroweak theory relies on "a neutrino which travels exactly with the velocity of light" [144]. Bilenky's review article restates the neutrino masslessness attribute of the electroweak theory: "Two-component left-handed massless neutrino fields play crucial role in the determination of the charged current structure of the Standard Model" (see the Abstract of [145]). Similarly, Srednicki states in his textbook: "Neutrino masses are exactly zero in the Standard Model" (see [146], p. 533). Contrary to this assumption, it is now recognized that the

neutrino is a massive particle [142].   In contrast, the DDWIT says that all weakly interacting elementary massive particles are Dirac particles.

2. The electroweak theory regards $W^\pm$ bosons as elementary particles. Unlike the case of the Dirac electron, it is proved in subsection 11.6.2, p. 184 that the electroweak theory cannot prove that its description of the $W^\pm$ bosons conserve charge. That is, the electroweak theory of the $W^\pm$ does not provide a conserved 4-current j^μ with a dimension of $[L^{-3}]$, and it satisfies the continuity equation $j^\mu_{,\mu} = 0$. Hence, the electroweak theory violates Maxwellian electrodynamics.   The DDWIT says that the $W^\pm$ particles are mesons of the top quark.   The data shown in table 14.1 on p.   220 strongly support the DDWIT.

3. The electroweak theory uses the Z boson as an elementary particle where the quantum function takes a mathematically real form.   The Z boson is a massive particle, and it was proved in section 4.2, p.   44, that a mathematically real function cannot describe a massive quantum particle. Hence, the mathematically real electroweak Z theory is wrong. The DDWIT says that the Z particle is a meson of the top quark.

4. A coherent quantum theory is based on a differential equation where the solutions appropriately describe relevant physical processes and states.   The Dirac theory of the electron is a good example of this issue.   By contrast, no electroweak textbook shows an *explicit form* of the required differential equation.   A fortiori, no solution of this unknown equation has been tested in terms of relevant experimental data.

5. An examination of DDWIT Lagrangian density (11.4) indicates that it would be interesting to briefly restate the contents of subsection 11.6.8.   The Lagrangian density (11.4) comprises just four terms, and its first three terms compose the Dirac Lagrangian density (5.17) and its electromagnetic interaction.   The last term, which represents weak interactions, takes the same general form as that of the electromagnetic interaction term: Each of these terms comprises a Lorentz scalar, which is a contraction of an external field with a tensor that depends on the Dirac γ matrices. This Lorentz scalar is multiplied by a factor that accounts for the strength of the interaction.

An important property of the DDWIT Lagrangian density (11.4) is that every interaction term is derivative-free. Hence, every interaction adds just one derivative-free term to the corresponding Hamiltonian. Therefore, the differential equation of the system is

$$i\frac{\partial \psi}{\partial t} = H\psi. \tag{11.23}$$

Here, the spatial derivatives of H are the ordinary derivatives of the Dirac equation.

The form of the electroweak Lagrangian density is completely different. It comprises more than 30 terms (see e.g. [47], p. 518; [48], pp. 113, 114; [49]). This unusual complexity is probably the reason why multiple SM textbooks take the strange approach of refraining from a presentation of the complete Lagrangian density of the electroweak theory.

The Occam's razor principle says that in the case of two competing theories with the same merit, one should prefer the simpler theory. In particular, it is stated above that if one theory relies on six assumptions while the other theory relies on more than thirty assumptions, then the Occam's razor criterion is decisive! This is an important argument that is added to the intrinsic errors of the electroweak theory mentioned above.

11.8 Three Kinds of Weak Interactions

The neutrino is a spin-1/2 massive particle that does not participate in strong and electromagnetic interactions. Therefore, this particle is useful for an examination of weak interaction properties. Experiments show that there are three kinds of neutrinos, called ν_e, ν_μ, and ν_τ [29]. Each of these neutrinos has an antiparticle.

The different names of the neutrinos indicate that they may behave differently in certain interactions. And indeed, "it was observed that the neutrinos produced from $\pi^+ \to \mu^+ \nu_\mu$ decays always produced a muon in charged-current weak interactions" (see [79], p. 329). Hence, the weak interaction term of the Lagrangian density

$$\mathcal{L}_{weak} = -\bar{\psi}\mathrm{d}\sigma_{\mu\nu}F^{\mu\nu}_{(w)}\psi. \tag{11.24}$$

does not describe all weak interaction details. The different behaviors of these neutrinos indicates that there are three kinds of weak interactions. These interactions apply separately to the different neutrino flavors and the three charged leptons. The CKM matrix indicates that analogous properties hold for the six quarks.

It can be stated that the Lagrangian density of 15.2 on p. 229 describes two kinds of electromagnetic-like interactions: One is charge dependent, and the other is monopole dependent. In contrast, there are three kinds of weak interactions, each of which depends on the flavor of the corresponding particle. Details of the actual structure of (11.24) and the interrelations between the three kinds of weak interactions are beyond the scope of this book.

11.9   Neutrinos in the Universe

Let us examine the neutrino's physical properties. Excluding gravitation, a neutrino participates only in weak interactions. The physical properties of weak interactions differ from the corresponding properties of strong and electromagnetic interactions. Parity violation and flavor violation are well-known properties of weak interaction. These effects enable the detection of a weak interaction process in events that are forbidden for strong and electromagnetic interactions, such as the β decay of nuclei, the decay of K mesons, and the decay of charged pions and muons.

Several kinds of experimental data show that the energy dependence of a weak process differs from that of electromagnetic and strong interaction processes as follows:

1. At low energy, weak interactions are really weak, and they are found *only* in a process that is forbidden for strong and electromagnetic interactions. Here, the time duration of a weak process is many orders of magnitude longer than that of a typical strong or electromagnetic process (see [23], p. 207). The lifetime of a process increases with the decrease of the strength of the relevant force. (The relative strength of the weak force under these circumstances is the reason for the name *weak interactions*.)

2. The energy dependence of the cross-section of a scattering process that is dominated by electromagnetic interactions is different from the case where the scattering process is dominated by weak interactions. Thus, we can compare the electron and the neutrino scattering data. The electron's electromagnetic cross-section decreases with the increase of energy (see [23], chapter 6), whereas the total cross-section of the weak interaction of neutrino scattering per nucleon increases

with energy (see [142], p. 1323). This means that at high enough energy, a weak interaction effect is expected to become quite powerful. (Here, the term *powerful* means strong, and its usage aims to avoid confusion with the ordinary strong interactions.) This argument uses neutrinos as pure sensors of weak interactions. However, the powerful attribute of high energy weak interactions applies to all Dirac particles.

3. The quark production $e^+e^- \to \bar{q}q$ of the electron-positron collision shows that at an energy of about 200 GeV, the relative strength of weak interactions is *stronger* than the electromagnetic interactions (see fig. 16.2 on p. 430 of [79]).

4. Let us examine the experimental results of proton-proton cross-section measurements [147]. The figure includes data of a large energy range, and the energy of a cosmic ray proton is about 10^5 times greater than that of the LHC, where the latter is the highest energy produced in laboratories. This means that there are regions in the universe where the energy of particle interaction is extremely high. The space just outside the event horizon of a black hole is a plausible candidate for this kind of region [148–150]. The previous arguments indicate that weak interaction processes are expected to take place in such a region. Referring to neutrino production, this process may be a direct $\bar{\nu}\nu$ pair production or a secondary neutrino production process, like that of the charged pion decay sequence [29]

$$\pi^- \to \mu^- \bar{\nu}_\mu$$
$$\mu^- \to e^- \bar{\nu}_e \nu_\mu. \tag{11.25}$$

There are other aspects of neutrino interactions. As stated in subsection 11.6.7, pair production and pair annihilation also apply to neutrino interactions. The data show that the neutrino mass is much smaller than that of the electron. Recent neutrino measurements indicate that the upper bound of a neutrino's mass is significantly smaller than 1 eV [151]; that is, it is smaller than 10^{-6} times the electronic mass. Moreover, each quark and lepton flavor participates in weak interactions. The tiny neutrino mass means that neutrino pair production is an effect that may take place in many scattering events.

Neutrino pair annihilation requires special examination. A neutrino does not participate in electromagnetic interactions. For example, the upper bound of the neutrino magnetic moment is smaller than 10^{-10} μ_B [29]. (This means that the strength of a $\bar{\nu}\nu$ magnetic interaction must be smaller by a factor of 10^{-20} with respect to the corresponding interaction of electrons.) Hence, a $\bar{\nu}\nu$ pair cannot directly decay into photons. It follows that a neutrino pair annihilation must go into a pair of Dirac particles. Therefore, ignoring higher order processes, a collision of two neutrinos where the invariant mass is less than $2M_e = 1.022$ MeV can only go to another pair (or pairs) of neutrinos. (M_e denotes the electronic mass.) This restriction means that at the ordinary region of the universe, the number of neutrinos with appropriately low energy does not decrease. In contrast, neutrino pair production may take place in a collision of cosmic rays of protons, electrons, and neutrinos with other neutrinos that already exist in the universe. This discussion indicates that, outside black holes, the number of neutrinos in the universe increases with time.

The tiny neutrino mass yields another attribute of this particle: Except at the very low energy region, its motion is extremely relativistic. Evidently, this neutrino velocity is larger than the escape velocity of cosmological bodies, except in places like the inner region of a black hole. Hence, free neutrinos are expected to be found in galactic and intergalactic space.

Experiments prove that a neutrino production effect is found in many natural processes of the present state of the universe. The following is a list of such events:

1. Many non-artificial nuclear isotopes undergo a β-decay. The decay of a free neutron

$$n \rightarrow p + e^- + \bar{\nu}_e \qquad (11.26)$$

is an example of such a process (see [8], p. 23). Another example is the decay of the carbon isotope

$$^{14}C \rightarrow {}^{14}N + e^- + \bar{\nu}_e. \qquad (11.27)$$

In general, every nuclear β-decay produces a neutrino or an antineutrino.

2. Pions are produced by an energetic collision, for example, by cosmic rays in the upper atmosphere. It is shown in (11.25) that neutrinos are emitted in the decay chain of a charged pion.

3. Nuclear reactions take place in stars like the sun. One result of these reactions is a neutrino emission (see [8], p. 366).

4. A supernova is a dramatic collapse of a star under its gravitational force. Here, protons eventually capture electrons and become neutrons. Neutrinos are emitted in this process (see [8], pp. 381, 382). The neutrinos emitted from the supernova 1987A that were detected on earth confirm this interpretation of a supernova process.

5. Galactic sources of high-energy neutrinos that are produced by binary stars where one of them is a compact object have been discussed [152, 153]. These systems are called microquasars.

6. Extremely high energy processes take place at the spatial region just outside the event horizon of a black hole. Particles' motion at this region is extremely relativistic, and energetic particle collision follows. In principle, pair production of all kinds of Dirac particles should be found in this region. The powerful property of weak interactions at extremely high energy mentioned above indicates that at this energy region, the neutrino pair production is not a negligible effect. Beyond this effect, pair production of particles like muons and charged pion production yield particles where the decay mode contains neutrinos (11.25). Furthermore, the tiny neutrino mass proves that relative to other massive particles, neutrinos are more likely to escape this region. These arguments explain why the outer part of a black hole is expected to be a "neutrino factory."

   It is interesting to note that the IceCube collaboration recently reported the detection of an extremely high energy neutrino emitted from a known blazar [154]. This neutrino may result from a direct pair production or from a decay sequence like that of a charged pion (11.25). In either case, it shows that very-high-energy neutrinos are produced in the outer region of a black hole.

7. The arguments outlined above mean that the neutrino population of the universe may be a non-negligible phenomenon. Therefore, high-energy cosmic rays of protons, heavier nuclei, and leptons may interact with these neutrinos and cause neutrino pair production events.

As explained above, a neutrino-antineutrino pair whose invariant mass is below 1.022 MeV cannot disappear. This means that low-energy neutrinos accumulate in the universe. Because of their extremely tiny mass, a considerable portion of these neutrinos moves relativistically. For this reason, they are evenly populated in galactic and intergalactic spaces. However, general considerations indicate that the density of extremely low-energy neutrinos is higher in the galactic inner region.

Astronomical observations indicate that galactic stars and their black holes cannot explain galactic gravitational phenomena. Referring to this issue, a review article states, "A general picture emerges, where both baryonic and non-baryonic dark matter is needed to explain current observations" [155]. One kind of missing matter is called *cold dark matter* (CDM). "CDM is thought to consist of particles (sometimes referred to as 'exotic' dark-matter particles) whose interactions with ordinary matter are so weak that they are seen primarily via their gravitational influence" (see [156] p. 306, [157]). A candidate for the missing matter is called the *weakly interacting massive particle* (WIMP) [29]. An experimental search for WIMPs has been carried out for several decades, but there is still no confirmation of the existence of relatively massive WIMPs.

Articles [155–157] identify neutrinos as possible candidates for a part of the missing mass of the universe. However, the arguments described above, which depend on the relatively new concept of massive neutrinos, change the picture. Considering a neutrino as an ordinary Dirac particle, the pair production effect and the blocked channel of low-energy neutrino pair annihilation indicate that the neutrino population of the universe may be quite significant and that neutrino's role should not be ignored.

Another astrophysical problem is the geometrical structure of the universe. This issue depends on the space-time curvature derived from a general relativistic treatment of the distribution of the entire energy/mass of the universe. It turns out that the global structure of the universe is still an open problem. For example, in a recently published article Di Valentino, Melchiorri, and Silk addressed this issue and argued that gravitational curvature renders a closed universe [158]. Another discussion of the structure of a closed universe can be found in the literature [159]. By contrast, other articles that have been published in the new millennium argue that the universe is flat [160, 161]. The discussion presented here describes new arguments that support the existence of galactic and intergalactic neutrino populations, and the gravitational field of these neutrinos may be used for the clarification of this open problem.

This section uses the relatively new evidence where neutrinos are massive Dirac particles. Neutrino pair production is one result of this issue. The data show that at very high energy the intensity of weak interactions becomes quite powerful. Hence, a neutrino pair production in regions of space that are close to the event horizon of a black hole is a significant process. Furthermore, the decay of particles like charged pions and muons produces neutrinos. Considering the other edge of the energy scale, it is explained above why the irrelevance of neutrino electromagnetic interaction means that for a low-energy neutrino collision, a neutrino pair annihilation can only go into a neutrino pair production. It can be concluded that outside black holes, the low energy neutrino population of the universe is likely to increase with time.

Pauli proposed the neutrino about 90 years ago, and its detectability has made progress since then. At present, energetic neutrinos can be detected by devices, whereas low-energy neutrinos are still undetectable in a direct sense. However, the theoretical arguments described above indicate the existence of low-energy neutrinos that are roaming elusively throughout the universe.

Astrophysical evidence indicates the existence of dark matter in the universe [156,157]. It turns out that the search for very massive WIMPs has not confirmed their existence [29]. However, if one takes the term WIMP literally, then a massive neutrino is a WIMP. The present analysis explains why neutrinos may be regarded as at least a part of the missing dark matter.

Chapter 12

Other Quantum Particles

This book discusses several kinds of elementary particles. They are massive Dirac particles: three kinds of neutrinos – ν_e, ν_μ, and ν_τ; three kinds of charged leptons – e, μ, and τ; six kinds of quarks – u, d, s, c, b and t; and their antiparticles. There is one kind of massless particle – the photon. The literature also shows an analysis of other kinds of elementary particles. Few remarks on these particles are presented in this chapter.

12.1   The $W^\pm$ and the Z Particles

The electroweak theory claims that the $W^\pm$ and the Z are elementary particles, are already disproved in chapter 11 of this book.

12.2   The Majorana Neutrino

Consider the Lagrangian density of a chargeless Dirac particle

$$\mathcal{L} = \bar{\psi}(i\gamma^\mu \partial_\mu - m)\psi, \tag{12.1}$$

and its Dirac equation

$$(i\gamma^\mu \partial_\mu - m)\psi = 0. \tag{12.2}$$

The definition of the Dirac γ^μ matrices is not unique. Majorana has contrived a set of pure imaginary γ matrices [162]; in his representation, the Dirac equation (12.2) is mathematically real. The

203

physical meaning of the Majorana equation is a neutrino that is identical to its antineutrino.

The following are two arguments that each refutes the Majorana idea:

1. It is already proved in this book (see section 4.2 on p. 44) that mathematically real equations cannot describe a massive quantum particle. Hence, a Majorana neutrino cannot exist.

2. This point examines a different aspect of the previous argument. As shown in section 4.2 on p. 44, a mathematically real wave function of a free massive particle moving along the x-direction takes the form

$$\psi(t, x) = A\sin(kx - \omega t - \delta), \qquad (12.3)$$

where A and δ are mathematically real quantities. Like the Dirac equation, the Majorana equation (12.2) is of the first order. Hence, the Majorana mass term of this equation contains the factor $\sin(kx - \omega t - \delta)$, whereas its derivative term contains the factor $\cos(kx - \omega t - \delta)$. This means that the Majorana function must vanish identically. This is a contradiction.

Experimental data support these theoretical arguments. An important fingerprint of the existence of a Majorana neutrino is the neutrinoless double β decay. Many experimental searches for this effect have been conducted, but the present conclusion says that "neutrinoless double beta decay has not yet been found" [163].

The order of magnitude of a typical half-life of the ordinary two-neutrino double β decay is 10^{21} years [164]. In contrast, the GERDA Collaboration has recently reported an upper bound of 10^{26} years for the neutrinoless double-beta decay of the ^{76}Ge isotope [165].

> Conclusion: The Majorana theory of the neutrino violates well-established physical laws. Therefore, a Majorana neutrino cannot exist. The failure of the experimental search for the neutrinoless double β decay supports these theoretical arguments.

12.3   The Klein-Gordon Equation

The KG equation was examined in the very early days of QM. A 1934 Pauli-Weisskopf article [141] showed many details of this equation. In particular, it provided an expression for the Lagrangian density that yields this equation. The KG equation describes two kinds of massive particles – a neutral particle and two particles that carry a charge $\pm e$, respectively. The mathematically real quantum function of the KG equation describes an electrically neutral particle, and the complex function describes charged particles. These quantum functions take the form $\phi(x)$, where x denotes the four space-time coordinates. This means that the KG particle is an elementary point-like particle. Many QFT textbooks treat the KG theory as a coherent physical theory and devote a full chapter to this equation.

The discoveries of the $\pi^{\pm}$ mesons in the 1940s and the π^0 meson in 1950 have been regarded as experimental support for the KG equation. Pions are classified as hadrons – namely, particles that participate in strong interactions. About twenty years later, quarks were recognized as the building blocks of hadrons. In particular, the three pions are regarded as bound $\bar{q}q$ states of the u and d quarks. This means that pions cannot be KG particles; this is simply because the KG function $\phi(x)$ depends on a single set of the four dimension coordinates, whereas a function that describes a bound $\bar{q}q$ state takes the form $\phi(\boldsymbol{x}_1, \boldsymbol{x}_2, t)$. The different number of independent coordinates of these functions proves this assertion. It has also been proved that the charge radius of the $\pi^{\pm}$ mesons is not much smaller than that of the proton [29]. Hence, the $\pi^{\pm}$ mesons are not elementary point-like particles, and none of them can be described by the KG equation. More than 80 years have elapsed since the publication of the Pauli-Weisskopf KG article, and the official publication of elementary particles [29] does not mention an experimentally established elementary particle that satisfies the Pauli-Weisskopf KG theory.

In contrast, the Dirac equation is a successful theory of massive quantum particles. This book agrees with the consensus that the three charged leptons e, μ, and τ are Dirac particles. Furthermore, it is now recognized that the neutrino is a massive particle. This book proves inconsistencies of the Majorana neutrino theory (see chapter 12.2 on p. 203). These inconsistencies and the failure to detect a Majorana neutrino support the claim that neutrinos are Dirac particles. Quarks are another kind of spin-1/2 particles. This book argues that the RCMT is the right strong interaction theory (see chapter 10 on p. 117). The strong interaction chapter of this book shows many failures of QCD, and these issues support the RCMT as the valid strong interaction theory. This evidence supports the RCMT claim that quarks are ordinary Dirac particles. All these points show how successful the Dirac equation is.

An important theme of this book is that mathematics plays a crucial role in the description of the physical world. Therefore, a question arises.

> *Question: Do mathematical flaws in the KG theory explain its systematic experimental failure?*

A "yes" answer is supported by the following list:

KG.1 It is proved in section 4.2, p. 44 that a mathematically real function cannot describe an elementary massive quantum particle. Hence, the mathematically real theory of a chargeless KG particle is wrong.

KG.2 The Lagrangian density of a charged KG particle is

$$\mathcal{L}_{KG} = g^{\mu\nu}(i\phi^*_{,\mu} + e\phi^* A_\mu)(i\phi_{,\nu} - eA_\nu\phi) - m^2\phi^*\phi \quad (12.4)$$

(see [141], p. 198). The interaction part of this expression has the term

$$\mathcal{L}_{Int\_2} = -e^2 g^{\mu\nu}\phi^* A_\mu A_\nu\phi. \quad (12.5)$$

This form is unacceptable because Maxwell equations are derived from a Lagrangian density that depends *linearly* on the 4-potential A_μ (see [3], pp. 78-80). Furthermore, the electromagnetic interaction term should be proportional to the strength e of the electric charge. The e^2 factor of (12.5) violates this requirement.

KG.3 The Noether expression for the 4-current of a charged KG particle is

$$j_\mu = i\phi_{,\mu}\phi^* - i\phi^*_{,\mu}\phi - 2eA_\mu\phi^*\phi \quad (12.6)$$

(see [141], p. 199). This 4-current depends on the external 4-potential. Again, like in (12.5), its application as an interaction term $ej^\mu A_\mu$ yields a Lagrangian density that depends quadratically on the 4-potential A_μ and the electric charge e as well. These properties are unacceptable.

KG.4 Let us examine the charge-free part of the KG 4-current (12.6). It comprises terms that contain a derivative of the KG function, and its density depends on the *time derivative*. Hence, in the Heisenberg picture, where the functions are time independent, the KG functions cannot be used for a basis of the Hilbert space. This means that the KG theory violates the required correspondence with QM.

Even one genuine contradiction justifies the rejection of a theory. We see that experimental results turn their back on the KG equation, and its theory is full of inherent contradictions that explain the failure to detect a genuine KG particle.

> *Conclusion: Although many SM textbooks ascribe physical meaning to the KG equation, both experiment and theory reject this equation.*

12.4 The Higgs Boson

The Higgs boson is an important element of the SM. According to the primary idea of this book, the Higgs Lagrangian density is analyzed. One term of the Higgs Lagrangian density is

$$\mathcal{L}_{Higgs} = m^2 \phi^* \phi + ... \tag{12.7}$$

(see e.g. [14], p. 715, [47], p. 515). The $[L^{-4}]$ dimension of the Lagrangian density proves that the dimension of the Higgs function ϕ is $[L^{-1}]$. In contrast, the dimension of the Dirac function ψ is $[L^{-3/2}]$ (see (5.17) on p. 62). These attributes are used later in this chapter. The term on the right side of (12.7) is derivative-free. Hence, apart from its sign, it takes the same form in the Higgs' Hamiltonian density. For this reason, the *energy of the Higgs particle depends quadratically on its mass*. This is inconsistent with the classical expression for the Lagrangian of a free particle that depends linearly on mass (see, e.g., (3.4) on p. 18).

Furthermore, The Higgs Lagrangian density (12.7) yields the Higgs energy-momentum tensor. Hence, the energy-density component of the Higgs energy-momentum tensor is

$$T^{00}_{Higgs} = -m^2 \phi^* \phi + ... \tag{12.8}$$

It means that for either a positive or negative Higgs mass $\pm m$, the energy density of the Higgs theory is *negative*. In other words:

> The Higgs theory says that Nature contains a particle
> whose energy density is negative and its entire energy
> is negative!

This result is totally inconsistent with experiment. It means that
the Higgs theory has no physical merits.

The previous argument demonstrates the powerful attribute of
well-established constraints that are imposed on physical theories.
However, I would guess that some readers are still unconvinced by
this argument. There is a rule saying that if a theory contains one
uncorrectable error then it probably contains some other errors of
this kind. With this assumption, let us examine other aspects of
the Higgs theory:

- Section 3.6 explains the vital role of partial differential equations as the equations of motion of an elementary quantum
particle. The theory of the Higgs particle is several decades
old. However, too many SM textbooks refrain from showing
an explicit form of the differential equation of the Higgs particle. A fortiori, no solution of this equation is shown to fit
the experimental data. In short, the Higgs theory violates the
constraint CNST.3 of section 3.9. These facts demonstrate
the dire state of the Higgs particle theoretical structure.

 This problematic situation can be illustrated as follows: Parts
 of the full Higgs Lagrangian density can be found in many
 places. One term of the Higgs Lagrangian density is

$$\mathcal{L}_H = -\lambda(\phi^\dagger\phi)^2 + ... \tag{12.9}$$

 (see e.g. [14], p. 715; [47], p. 515). The quartic term of
 (12.9) means that the Higgs equation of motion is *not linear!*
 In contrast, ordinary quantum equations are linear. Linearity
 is the underlying property that explains the interference of
 waves, the relevance of the Hilbert space, and many other
 intrinsic properties of quantum theories.

 > *Conclusion: The Higgs equations of motion are yet
 > unknown. Therefore, the Higgs theory is alien to the
 > standard structure of quantum theories.*

- Another term of the Higgs Lagrangian density is

$$\mathcal{L}_H = \phi^\dagger_{,\mu}\phi_{,\nu}g^{\mu\nu} + ... \tag{12.10}$$

(see e.g. [14], p. 715; [47], p. 515). The right side of (12.10) agrees with an analogous term of the KG equation. This term contributes a *symmetric* quantity (with respect to $\phi^\dagger, \phi$) to the Higgs Hamiltonian density (see. e.g. [141], p. 192). In contrast, the 4-current of this term and its associated density are *antisymmetric* with respect to $\phi^\dagger, \phi$ (see. e.g. [141], p. 193). Hence, unlike the case of a Dirac particle, it is impossible to extract the Hamiltonian operator from the Hamiltonian density. The KG equation faces the same problem. And indeed, many decades have elapsed and textbooks still do not show a coherent expression for the *Hamiltonian operator* of the KG particle or the Higgs particle.

Without a Hamiltonian operator, the Higgs theory cannot construct a coherent Hilbert space and calculate appropriate quantities. In particular, the $[L^{-1}]$ dimension of the Higgs function proves that substitution of the Higgs function ϕ into the standard definition of the Hilbert space inner product

$$< \phi^\dagger | \phi > = \int \phi^\dagger \phi \, d^3 r, \qquad (12.11)$$

is dimensionally inconsistent (see e.g. 3.15 on p. 24). This means that the standard route towards a coherent Hilbert space is blocked for the Higgs theory. This is another violation of the correspondence principle.

> *Conclusion: The theory of the Higgs particle is inherently wrong because it violates fundamental requirements that are relevant to a theory of an elementary quantum particle.*

SM supporters have declared that the CERN 125 GeV particle is a Higgs boson. This issue belongs to the experimental realm, and it is discussed in section 14.1.2.

12.5   The Proca Photon

The Proca Lagrangian density that describes a massive photon takes the form

$$\mathcal{L}_{Proca} = -\frac{1}{16\pi} F_{\mu\nu} F^{\mu\nu} + \frac{m^2}{8\pi} A_\mu A^\mu - j^\mu A_\mu, \qquad (12.12)$$

where m denotes the photon's mass (see [28], pp. 597-601). The second term of (12.12) indicates that it is a modification of the ordinary electromagnetic Lagrangian density (see [3], p. 78). The Proca Lagrangian density (12.12) justifies its discussion in this chapter, which discusses quantum particles that are based on an appropriate Lagrangian density.

The fields $F^{\mu\nu}$ and the potentials A_μ are mathematically real quantities. It was proved (see section 4.2 on p. 44) that a mathematically real function cannot describe a massive quantum particle. This means that the Proca idea of a massive photon is theoretically unacceptable.

The experimental side supports the masslessness of the photon. Thus, [29] said that the upper bound of the photon's mass is 10^{-18} eV. This means that the experimental upper bound of the photon's mass is smaller than 10^{-23} times the electron's mass!

Conclusion: Theory and experiment refute the Proca idea of a massive photon.

Chapter 13

Miscellaneous Ideas

Contemporary physics contains theoretical ideas that are based on the concept that the SM is a good theory but not a perfect one. Hence, it should be extended to improve the structure of theoretical physics. The term *beyond the SM* describes this set of theories. The scientific work that follows this approach is intensive. Today (August 2021), Google says that the string of words "beyond the standard model" occurs on the web more than 6 million times! Some theories that belong to this set have yielded predictions that can be tested experimentally. This chapter examines theories that belong to this set, as well as some other physical ideas that deviate from the SM. As of today (August 2021), experiments do not support these ideas.

This book describes abundant SM errors. Therefore, it is clear why it does not follow this approach. *In contrast, it argues that the primary task of contemporary theoretical physics is not to extend the SM but to replace it with correct theories!*

13.1 Supersymmetry

Supersymmetry (SUSY) is a theoretical idea that belongs to theories that are called "beyond the SM." A basic SUSY attribute is that nature comprises particles that can be organized in pairs. Each fermion has a boson counterpart and vice versa. The standard SUSY terminology adds a prefix s to the name of every SUSY counterpart of a known particle. For example, the electron's SUSY counterpart is called a selectron.

This book denies SUSY for the two following reasons:

1. The general argument says that since the SM is plagued with plenty of errors, its extension is doomed to add further errors.

2. The concept of an elementary particle is an essential feature of this book. Quantum functions that are used in the Lagrangian density of a quantum theory describe elementary particles. The genuine elementary massive particles of this book are the three charged leptons (electron, μ, and τ); the three corresponding neutrinos (ν_e, ν_μ, and ν_τ); the six quarks (u, d, s, c, b, t); and their 12 antiparticles. There is one kind of massless particle, which is the photon.

   Experiments show that the above-mentioned particles behave as elementary point-like particles, as listed in [29]. This book proves that the SM interpretation of the $W^\pm$, Z, and the Higgs 125 GeV bosons is wrong. Furthermore, it proves that massive particles are Dirac particles. Another point of this book is that theories of a massive boson that can be found in the literature, are wrong. Hence, to be considered seriously, SUSY should show a coherent theory of a massive boson and its interaction.

As of today, the LEP Collider, the Tevatron, and the LHC have tried to detect SUSY particles [166]. All these attempts have ended in vain, and the most recent review states: "There is no experimental evidence that a supersymmetric extension to the Standard Model is correct" [166].

> *Conclusions: This section has explained why SUSY does not take the right course needed for the progress of physics. Experiments support this conclusion, and all attempts aiming to detect SUSY particles have failed.*

13.2   Grand Unified Theory

A grand unified theory (GUT) is a theoretical idea stating that at very high energy, the three relevant interactions – strong, electromagnetic, and weak – merge into a single force [167]. As stated in many places of this book, well-established experimental data play an important role in the structure of theoretical physics. Furthermore, mathematics is another important element that affects the structure of theoretical physics. Relying on these issues, it is briefly explained here why the GUT concept does not promise to be the right direction of the progress of physics.

Table 9.1 on p. 115 shows data on the forces mentioned above. Evidently, strong and electromagnetic forces have something in common, whereas weak interactions have completely different properties. This experimental information affects the structure of the interaction term that describes how Dirac particles (quarks, charged leptons, and neutrinos) interact with appropriate fields. The strong and the electromagnetic interaction terms of the particle's Lagrangian density take the form of a contraction of the Dirac 4-current with an appropriate 4-potential. In contrast, the weak interaction term takes the form of the Pauli term, where the particle's spin $\sigma_{\mu\nu}$ is contracted with an external second rank antisymmetric tensor (see section 15.2 on p. 229). The dimension of the 4-potential is $[L^{-1}]$, and the dimension of the weak interaction fields' tensor mentioned above is $[L^{-2}]$.

The increase of the energy scale is a continuous process, whereas neither dimension nor tensorial rank vary continuously. Also parity conservation does not vary continuously. Because of the intrinsic difference between their interaction terms, weak interactions cannot merge with strong or electromagnetic interactions. This argument refutes the primary GUT predictions.

The proton decay idea is derived from GUT. Experimental attempts to detect this effect have yielded a null result (see section 13.3).

> *Conclusions: Contrary to the primary GUT idea, such a continuous process as the increase of energy cannot induce weak interactions to merge with strong and electromagnetic interactions. Furthermore, experimental evidence does not support a GUT.*

13.3   The Proton Decay Idea

The proton decay idea is derived from the GUT extension of the SM. This idea violates a fundamental principle called conservation of baryonic number. Considering the inherent errors of the GUT, proton decay is regarded here as an erroneous idea. Experimental efforts aiming to detect proton decay have ended in vain. For example, the present (August 2021) Wikipedia entry on proton decay states: "To date, all attempts to observe these events have failed" [168].

This book supports the idea that even if an experimental work is based on an erroneous concept, it may improve many aspects of physics. It can be said that experimental work enhances the skill of experimenters in general and that of new experimenters in particular. Even if the work ends in vain, it is important to show that experiments do not support a given physical idea. Furthermore, experiments may accidentally detect new physical effects.

An example that illustrates this opinion is as follows: Consider the experimental work dedicated to the detection of a proton decay event. To identify this event, detectors were built in old mines deep below the Earth's surface in several places around the globe. One result of these experiments shows how the apparently erroneous idea of "proton decay" has made a positive contribution to physics.

About 168,000 years ago a supernova exploded at the Large Magellanic Cloud. This explosion produced a gigantic amount of neutrinos in a short time. Hence, a thin spherical shell of neutrinos propagated throughout space at a speed very close to the speed of light. Table 13.1 shows dates of historical landmarks of human activity that are relevant to the detection of these neutrinos. Its first column shows the event's year; the second shows the number of years left from the event's date until the 1987A supernova explosion was detected on planet earth. The third column shows

Table 13.1: Historical events.

Year	Years left	Percents	Event
~-166000	~168000	0	The 1987A supernova explosion
~-10000	~12000	93	End of the last ice age
17th century	~300	99.8	Newton – the beginning of physics
1905	82	99.95	The birth of SR
~1926	~60	99.96	The birth of QM
1930	57	99.96	Pauli proposes the neutrino
1956	31	99.98	The first neutrino detection
1967	20	99.99	The proton decay suggestion
1970+	~15	99.99	Work for the proton decay detection
1987	0	100	Boom! ~25 neutrinos recorded within ~15 sec by devices built for the proton decay detection

the percentage of the period between the explosion date and the event's time respecting the period between the explosion date and the time of its detection on planet Earth. The table's data show that the neutrino detection took place at virtually the last moment.

Supernova theories show that these events produce an abundant number of neutrinos. The number of measured neutrinos of the supernova 1987A, their energy distribution, and the time interval of their detection may support the corresponding theoretical work. Hence, the erroneous idea of proton decay enabled the neutrino detection of the 1987A supernova.

13.4 Gravity

It was stated near the beginning of this book that it discusses strong, electromagnetic, and weak interactions. Gravitational interaction is outside its scope. The SM discusses the same topics. Several arguments explain why gravitation is a different interaction that justifies its exclusion:

G.1 *A different mathematical structure.* The theoretical side of this book examines elementary point-like quantum particles. The primary expression is the Lagrangian density yielding the equations of motion that determine the time evolution of the system. The most important expression of this book shows the Lagrangian density of a spin-1/2 Dirac particle. It comprises terms of its strong, electromagnetic, and weak interactions. The structure of this expression is already known to readers of this book. For the convenience of the discussion, this expression is copied here:

$$\mathcal{L}_{full} = \bar{\psi}[\gamma^\mu i\partial_\mu - m - e\gamma^\mu A_{(e)\mu} - g\gamma^\mu A_{(m)\mu} - \mathrm{d}\sigma_{\mu\nu}F^{\mu\nu}_{(w)}]\psi. \tag{15.2}$$

The third, fourth, and fifth terms on the right side of this expression represent the electromagnetic, strong, and weak interactions, respectively. The electromagnetic interaction is the ordinary QED term that can be found in many textbooks. In (15.2), terms that represent strong and weak interactions take an analogous form of a Lorentz contraction of Dirac γ matrices with an external field.

The case of the gravitational interaction is completely different:

A. The Einstein equations of the gravitational field depend on the energy-momentum tensor of matter (see [3], p. 297). The entire Lagrangian density of a coherent field theory provides this tensor (see chapter 5).

B. In contrast to the interaction terms of 15.2, gravitation affects the motion of particles via the change of the metric of a flat space-time into that of a curved space-time. Here, the Lorentz uniform diagonal metric (1,-1,-1,-1) changes, and the metric $g_{\mu\nu}$ is obtained from a solution of the Einstein equations of the gravitational field. For example, the electromagnetic interaction term is a contraction $-e\gamma^\mu A^\nu g_{\mu\nu}$ that uses the metric.

It is shown above that the relation between quantum theories of elementary particles and general relativity is not the same as that of QFT interactions. Gravitation does not enter by means of a specific interaction term; instead, it changes the metric, which is used in every term of the equation.

> *Conclusion: Gravitational interaction does not fit the mathematical framework of the three relevant interactions shown in (15.2).*

G.2 This book supports the assertion that elementary massive particles satisfy the Dirac equation. This equation is linear. Furthermore, the mediating fields satisfy the Maxwell equations, which are also linear. In contrast, the gravitational field equations are non-linear. The different structure of the fields' equations indicates an inherent difference between quantum theories and gravitation.

G.3 Einstein's explanation of the photoelectric effect, as well as the Compton scattering, prove that electromagnetic radiation is an assembly of photons. Laboratory measurements substantiate the photon theory of electromagnetic radiation. There is no analogous experiment for the radiation of gravitational field showing that it comprises gravitons. Furthermore, because of the extremely weak gravitational interaction, it can be stated with relative certainty that a measurable gravitational radiation will never be produced in laboratories.

> *Conclusion: There is no experimental support for the graviton existence.*

G.4 In experiments carried out in Earth laboratories, quantum processes are not affected by gravitational interaction. Here the strong, electromagnetic, and weak interactions are stronger than the gravitational interactions by dozens of orders of magnitude. Hence, ignoring gravitation is irrelevant to the important test of the three theories – their fit to experimental data.

The above arguments explain the scope of this book and clarify why gravitational interactions are excluded.

Chapter 14

An Unbiased Data Observation

This chapter presents experimental data that indicate systematic attributes of particles.

14.1 The Problem with the $W^\pm$, the Z, and the Higgs particles

Chapter 11 proved many erroneous elements of the electroweak theory. One result of this evidence is that the electroweak explanations of the $W^\pm$ and the Z particles are unacceptable. Analogously, section 12.4 on p. 207 proved errors of the Higgs theory. Hence, the present interpretation of the H^0 particle is also unacceptable.

Two problems follow these issues:

P.1 What is the physical meaning of the mentioned particles?

P.2 Where are the top quark mesons?

SM supporters say that "because, as we now realize, the top quark is simply too short-lived to form bound states – apparently there *are* no top baryons and mesons" (See [67], p. 47). This argument is quite strange, and it indicates a typical wishful thinking explanation. Indeed, if the top quark is simply too short lived, its bound states should also be short lived! Here is an argument that shows another aspect of this point. Width of 2 GeV would completely obliterate the peak of the ρ meson whose mass is about

750 MeV. However, such a width is found and measured for particles whose mass is around 100 GeV. Hence, a meson whose mass is around 100 GeV and its width is around 2 GeV is a detectable particle. Here weak interactions are powerful, the meson's lifetime is short, and its associate energy-width is much larger than the width of light mesons.

Physics is an experimental science, and the answer to problems P.1, P.2 is derived from an analysis of the data. This chapter undertakes this task.

14.1.1   Similarities

The Top and the W Decay Similarity.

The top quark and the $W^{\pm}$ particles are electrically charged objects. Therefore, their decay products must comprise an electrically charged particle. The RCMT says that $W^{\pm}$ particles are mesons of the top quark. This means that similarity between the properties of these particles is likely to be found. In contrast, the SM says that the top quark is an elementary particle that is *affected* by strong, electromagnetic, and weak interactions, whereas $W^{\pm}$ are elementary particles that *mediate* weak interactions. Therefore, the SM says that the top quark and the $W^{\pm}$ are inherently different particles. Hence, any similarity between their properties is just an accidental effect (or a miracle...).

Table 14.1: The decay modes (in %) of the top quark and the W^{+} [29]

Channel	top quark	W^{+}
$\nu_e + X$	11	11
$\nu_\mu + X$	11	11
$\nu_\tau + X$	11	11
hadrons	67	67

Table 14.1 shows the percentage (rounded to two decimal digits) of the neutrino and pure hadronic decay channels of these particles. No one can deny the striking similarity in the data. The arguments mentioned above mean that this strong similarity between the top quark and the $W^{\pm}$ particle supports RCMT. In contrast, SM supporters apparently believe in miracles.

The Width Similarity.

The decay of unstable particles induces a limited indeterminacy in their lifetime. The uncertainty principle means that the energy

Table 14.2: Mass and energy width of heavy particles (in GeV) [29]

Particle	Mass	Width
Top	172.7	1.42
H^0	125	?
Z	91.2	2.50
$W^{\pm}$	80.4	2.01

of these particles is not well defined. For practical reasons, scientific data show the half-life of long-lived particles, whereas the energy width is given for rapidly decaying particles. Table 14.2 shows the energy width of the four heaviest particles. The energy-width of these particles is large and similar. (Note that at present the official opinion says that the total decay width for a light Higgs boson with a mass in the observed range is not expected to be directly observable at the LHC.)

Remarks:

- Officially, the width of the H^0 is still undetermined. Let us wait and see data from future electron-positron colliders. The RCMT says that the H^0 is a $\bar{t}t$ meson. Hence, its width is expected to be similar to the values of the other particles that are reported in table 14.2.

- The RCMT says that the $W^{\pm}$, Z particles are mesons of the top quarks. The top quark decays rapidly, and this attribute is the reason for the fast decay of the $W^{\pm}$, Z particles. Hence, the energy width of these particles is similar. In contrast, the SM says that the top quark is an elementary particle that is *affected* by external interactions, while the $W^{\pm}$, Z particles are point-like elementary particles that *mediate* the weak interactions. The similar width that is shown in table 14.2 supports the RCMT.

14.1.2 The e^+e^- Scattering Experiments

Figure 10.17 on p. 262 of [79] shows the energy dependence of the ratio between the hadronic and the leptonic cross-sections of e^+e^- collisions at an energy below 35 GeV. The figure shows several peaks around 1 GeV; they belong to the ρ^0, ω, and ϕ mesons. These mesons are states of $\bar{u}u$, $\bar{d}d$, or $\bar{s}s$ quarks. Two strong peaks around 3 GeV pertain to the $\bar{c}c$ mesons. Several peaks of $\bar{b}b$ mesons are found around 10 GeV. Goldhaber and Wiss presented a more detailed description of the e^+e^- collisions data at this energy range [169].

The e^+e^- meson production at energies below ~ 10 GeV is determined by the electromagnetic interaction of the colliding particles, and the process conserves flavor and parity. For this reason, only quark-antiquark pairs of the same flavor are produced.

There is information about more energetic e^+e^- collisions. An impressive peak that represents the Z particle is seen around 91 GeV (see [14], p. 711). The energy dependence of weak interactions is discussed in section 11.1 on p. 171. As shown there, the experimental data prove that weak interactions dominate processes at the Z energy. Hence, flavor violation processes like the production of $t\bar{c}$, $t\bar{u}$, or their antiparticle counterparts, may take place. To estimate these states, let us examine some other meson data.

Table 14.3 shows relevant data of pairs of mesons. The second column shows the meson's quark components. (Note that fluctuations like $b\bar{s} \leftrightarrow \bar{b}s$ are irrelevant to the present discussion.) The third column shows the meson's mass, and the last column shows the mass difference between these mesons. This table shows data of three pairs of mesons. In each pair, an $\bar{s}$ quark replaces a $\bar{d}$ quark. In the case of K^+, π^+, the difference between the masses is significant. In contrast, the situation of the heavier pairs is different. The data show that the mass difference between the mesons decreases with the increase of the meson's mass. Relying on this evidence, one expects that the mass difference between the $t\bar{c}$ and the $t\bar{u}$ mesons will be small in relation to the width of the Z particle, which is about 2 GeV. These arguments lead to the conclusion that the Z particle is a bound state of two kinds of quark pairs, $t\bar{c}$ and $t\bar{u}$, or their antiparticle counterparts. The primary binding force is the weak force.

An analogous argument holds for the $W^\pm$ particles. Furthermore, the top quark is much heavier than every other quark. Hence,

Table 14.3: Mass Data (in MeV) of Pairs of Mesons

Name	Quarks	Mass	ΔM
K^+	$u\bar{s}$	494	354
π^+	$u\bar{d}$	140	
D_s^+	$c\bar{s}$	1968	98
D^+	$c\bar{d}$	1870	
B_s^0	$b\bar{s}$	5367	87
B^0	$b\bar{d}$	5280	

the $\bar{t}t$ meson is expected to be heavier than either the $W^{\pm}$ or the Z. This is the particle of 125 GeV that is called the Higgs boson.

> *Conclusion: Experimental data are compatible with the claim that the states of the $W^{\pm}$, Z, and H^0 particles are mesons of the top quark.*

Chapter 15

Concluding Remarks

This chapter briefly describes some important points that have been discussed in this book. This presentation aims to help readers see the full picture of a coherent structure of particle physics theories. The arguments emphasize the central role of the variational principle and the Dirac theory of elementary massive particles in the theoretical framework of particle physics.

15.1   The Merits of the Variational Principle

At present (August 2021), mainstream physicists regard the SM as an assembly of correct theories. This book disagrees with many SM elements. However, it is important to point out that two fundamental SM elements are agreed on:

1. This book agrees with the SM on the definition of the domain of physical phenomena that should be examined. The present Wikipedia entry states [170]: "The Standard Model of particle physics is the theory describing three of the four known fundamental forces (the electromagnetic, weak, and strong interactions, and not including the gravitational force) in the universe..." The present book is dedicated to physical phenomena that belong to this domain.

2. Textbooks on QFT aim to describe details of the SM (see e.g. [14, 20, 46]). These textbooks use the variational principle as the basis for a QFT of a given elementary particle. Each specific interaction that applies to a given elementary

particle has a Lagrangian density. This Lagrangian density is regarded as the theoretical cornerstone of the interacting particle. This book agrees with this approach but not with the detailed form of the SM Lagrangian density of the presently known elementary particles and their interactions.

Physics is a mature science, and every theory should abide by relevant principles. These principles impose constraints, and any theory that violates even one of these constraints should be rejected. However, it is a good idea to reexamine the validity of the constraints every once in a while.

A description of the constraints that this book adopts were listed in section 3.9 that begins on p. 39. Each constraint is regarded here as a crucial element of a quantum theory of an elementary particle. Other valid constraints are derived from them. They are restated here in a different form:

Req.I The theory should be derivable from a Lagrangian density.

Req.II The theory should abide by SR.

Solution: The Lagrangian density should be a Lorentz scalar.

Req.III Specific terms of the Lagrangian density represent each of the three relevant interactions.

Req.IV A partial differential equation is a crucial element of a theory of an elementary particle. This equation is the Euler-Lagrange equation, derived from the Lagrangian density of the particle's theory.

Req.V Solutions of the theory's equations of motion must describe the states and the time evolution of the relevant elementary quantum particle.

Req.VI A theory of a massive quantum particle should provide a coherent expression for the wave equation, and the limit of its free particle solution should agree with the de Broglie principle.

Solution: The Euler-Lagrange equation of the variational principle should take the form of a wave equation. The action is the phase, and its free-particle limit should agree with the de Broglie principle. The action should be a mathematically real dimensionless Lorentz scalar. For this reason, the Lagrangian density must be a mathematically real Lorentz scalar whose dimension is $[L^{-4}]$.

Req.VII The appropriate limit of quantities of a higher rank theory should be compatible with corresponding quantities of a lower rank theory. Hence, the appropriate limit of quantities of a QFT of a massive particle should agree with corresponding quantities of QM (see [20], p. 49).

Req.VIII An elementary particle is point-like.

Solution: The theory's Lagrangian density should depend on quantum functions with the form $\psi(x)$, where x denotes a set of the four space-time coordinates.

Req.IX The theory must conserve energy, momentum, and angular momentum.

Solution: The Lagrangian density should not depend explicitly on the space-time coordinates. In this case, the Noether theorem yields the theory's energy-momentum tensor.

Req.X A theory of a massive quantum particle should provide a coherent expression for density.

Solution: The Noether theorem provides this kind of expression. A coherent expression for density should not depend on derivatives of the quantum functions with respect to the space-time coordinates.

Req.XI The interaction term of the Lagrangian density of a charged particle should be proportional to its electric charge e.

Req.XII Maxwell equations are independent of the 4-potential A_μ. In VE, Maxwellian electrodynamics uses A_μ as the coordinate of the electromagnetic Lagrangian density (see [3], section 30). Therefore, the Euler-Lagrange equations prove that no term of the Lagrangian density should have A_μ whose power is greater than unity.

Req.XIII A theory of a quantum particle should abide by Wigner's classification of physical particles: There are two sets of physically meaningful particles – massive particles that have positive mass and spin, and massless particles that have positive energy and two degrees of helicity [16].

This book adheres to the idea that elementary massive particles are spin-1/2 particles that are described by the Dirac equation. (For details, see the next section. Experiments have shown that this set of particles comprises the following:

- Three flavors of neutrinos: ν_e, ν_μ, ν_τ.

- Three flavors of charged leptons: e, μ, τ.

- Six flavors of quarks: u, d, s, c, b, t.

Each of these elementary particles has an antiparticle.

It is proved in this book that the SM theoretical description of the particles called $W^\pm$, Z, and Higgs bosons contains erroneous elements. Hence, these particles are not elementary point-like particles, but instead, mesons of the top quark.

Some textbooks put forward the problem of whether the variational principle is a mandatory theoretical element of QFT. For example, Weinberg says: "If we discovered a quantum field theory that led to a physically satisfactory S-matrix, would it bother us if it could not be derived by the canonical quantization of some Lagrangian?" (see [20], p. 292). However, in this same textbook Weinberg examines a Lagrangian density of the form

$$\mathcal{L}[\psi(x), \psi(x)_{,\mu}] \tag{15.1}$$

and its associated Lagrangian, and states (see p. 300): "All field theories used in current theories of elementary particles have Lagrangians of this form." This book adopts the form of (15.1).

The following arguments explain why this book evades the problem of the possibility of constructing a coherent physical theory that cannot be derived from the variational principle:

- The variational principle has a magnificent mathematical tool called the Noether theorem. This theorem instructs people on how to build a Lagrangian density that abides by many of the requirements Req.IV – Req.XIII.

- This book aims to solve the problems of *existing particles*. As stated above, it explains why all elementary massive particles are spin-1/2 Dirac particles.

- It is well known that the Dirac equation can be derived from a Lagrangian density. Therefore, there is no indispensable need for a theory that takes a different form.

- The long list of requirements Req.IV – Req.XIII explains why the construction of a coherent theory of an elementary particle that *cannot* be derived from a Lagrangian density is a very difficult assignment.

- This book describes a novel application of the Noether theorem. It can be applied to a given Lagrangian density and discover erroneous theoretical elements (see chapters 5, 6). Evidently, error removal is a vital element of any scientific work.

15.2 Theory of an Interacting Dirac Particle

As explained in the previous section, this book argues that the Lagrangian density is the cornerstone of a quantum theory that describes the state and the time evolution of a quantum particle. Let us see the complete Lagrangian density of a spin-1/2 Dirac particle and its interactions:

$$\mathcal{L}_{full} = \bar{\psi}[\gamma^\mu i\partial_\mu - m - e\gamma^\mu A_{(e)\mu} - g\gamma^\mu A_{(m)\mu} - \mathrm{d}\sigma_{\mu\nu}\mathcal{F}^{\mu\nu}_{(w)}]\psi. \quad (15.2)$$

(The self term of the electromagnetic fields $-\frac{1}{16\pi}F_{\mu\nu}F^{\mu\nu}$ is omitted from (15.2)). Each term of this Lagrangian density is a Lorentz scalar with a dimension of $[L^{-4}]$. The first term inside the square brackets is the ordinary kinetic term, the second is the mass term, the third is the QED interaction term, the fourth is the strong interaction term, and the last is the weak interaction term. Here $\mathcal{F}^{\mu\nu}_{(w)}$ denotes the field of weak interactions. For a neutrino, e=g=0, and only the terms 1, 2, and 5 of (15.2) hold. For a charged lepton, $g = 0$, and the fourth term does not hold. All terms of (15.2) apply to quarks.

Each interaction term of (15.2) comprises three factors. The first factor denotes the strength of the interaction between the Dirac particle and the external field. The second factor comprises the γ^μ 4-vector or the antisymmetric product $\sigma_{\mu\nu}$ of two γ_μ. Finally, the last factor is a tensorial expression of the external field. The γ^μ matrices are a crucial mathematical asset of the Dirac theory. They are numerical dimensionless 4-vectors, and they (or their antisymmetric product) may contract with a tensorial expression of an external field and produce the required Lorentz scalar *without changing the dimension of the term*. In particular, every interaction term does not contain the derivative operator ∂_μ. Furthermore, the Noether expression for the 4-current (see 3.14 on p. 24) depends on the derivative of the quantum function ψ_μ. Hence, like in the well-known electromagnetic interaction of a Dirac particle, the absence of the derivative operator from the interaction terms of

(15.2) is a very important aspect of the Dirac theory: *it means that the interaction term and the Noether theorem for the 4-current do not modify the original expression for density of the Dirac particle.*

The literature calls the electromagnetic interaction of (15.2) the *minimal interaction* (see, e.g., [2], p. 11). It was shown above that every interaction term of (15.2) has an analogous structure. It is just one term that is a product of three factors. For this reason, one may argue that every interaction term of (15.2) is a *minimal interaction*. Another argument that justifies this conclusion goes as follows: Each of the three relevant interactions has specific physical properties. Hence, the associated interaction term of the Lagrangian density should have a unique form. Therefore, each of the three relevant interactions should have at least one distinct interaction term. Hence, remembering that the metric is used for the gravitation interaction, (15.2) is a *minimal interaction* because just one term is used for every interaction. A justification for this assertion relies on a comparison with the excessive number of terms used by the SM for its description of the strong and weak interactions.

Indeed, the SM strong interaction sector is called QCD. This theory is certainly more complicated than that of (15.2) because it adds another degree of freedom, called color. Color is an extension of the charge concept and together with its field they belong to the non-commutative SU(3) group. QCD also relies on a hypothesis that forbids a colored structure of its quark to exist as a free particle. As proved in chapter 10, QCD has been constructed on an erroneous basis, and it is inconsistent with many kinds of experimental data.

The case of the SM weak interaction sector, called the electroweak theory, is even worse than QCD. Its Lagrangian density has more than 30 terms!!! (See subsection 11.6.8). Chapter 11 pointed out many problematic electroweak issues. For example, its description of the electrically charged particles called $W^{\pm}$ violates Maxwellian electrodynamics.

Finally, the general criterion called Occam's razor certainly favors the Lagrangian density (15.2) over the multitude of terms of the SM version of these interactions. Moreover, as stated in the Occam's razor section (see section 2.2, p. 9): "if one theory relies on six assumptions while the other theory relies on more than thirty assumptions, then the Occam's razor criterion is decisive!"

> *Conclusion: The Occam's razor principle strongly supports the theories of this book (15.2) and denies the SM in general and its electroweak theory in particular.*

15.3 The Significance of SM Errors

This book describes fundamental elements of theoretical physics of the strong, electromagnetic, and weak interactions. The SM is the currently accepted theory of these sectors. This book proves that too many errors are present in the SM.

Most people regard error correction as an important assignment. The author of this book agrees with this opinion. Correcting SM errors is even more urgent because too many people and institutes unjustifiably praise SM. The following are just four glorifying examples, two of which are taken from textbooks and the other two from publications of large research institutes:

PRAISE.1 "At various points in our discussion, we have noted that these theories have passed stringent quantitative experimental tests." ("these theories" == SM) (see [14], p. 781).

PRAISE.2 "Remarkably, the Standard Model provides a successful description of all current experimental data and represents one of the triumphs of modern physics" (see [79], p. 1).

PRAISE.3 On January 7, 2015, CERN announced:

"The standard model describes everything we know about the smallest building blocks of nature yet observed. It's the most accurate theory ever developed, in any field" [171].

PRAISE.4 An official Fermilab publication declares on November 18, 2011:

"The Standard Model: The most successful theory ever" [172].

This declaration is repeated on December 18, 2016 [173].

Many other SM glorifying statements can be found in the literature and on the web.

Most people are amazed by the achievements of modern technology, and they know that physics provides a strong basis for many elements of this technology. The success of this technology and the reliability of its products provide a good reason for people to believe that physics is a correct and accurate science. However, physics is an exact science that is based on an adequate description of experimental data and a coherent mathematical structure. Public opinion in general, and physicists' opinions in particular, are irrelevant to the veracity of a physical theory. This book adheres to the search for scientific truth and ignores sociological issues that lie outside the domain of physics.

The systematic cover-up of many erroneous SM elements is possible because the SM has practically nothing to do with the present technology. As a matter of fact, this technology uses well-established laws of electrodynamics and QM. Other SM elements are irrelevant. Simple examples that demonstrate the dubious SM structure are as follows: An important QED process is called *renormalization,* and Feynman – a key person in QED development – said that renormalization is "a dippy process" (see [77], p. 128). How can "the most successful theory ever" be based on a dippy process? Furthermore, the "Proton Spin Crisis" is an unsettled problem. How can "the most successful theory ever" have such a problem? The next sections of this summary chapter give lists of SM errors that are discussed in this book and references to a more detailed discussion of their meaning.

15.4   Errors in Quantum Theories

The following items provide a summary of errors of the present structure of quantum theories.

1. Mathematically real quantum functions cannot describe an elementary massive quantum particle (see section 7.3 on p. 76.)

   Conclusions: The electroweak description of the Z particle and the mathematically real version of the Higgs particle are wrong. The Majorana neutrino theory is wrong, and the Proca theory of a massive photon is wrong. The mathematically real version of the KG theory is wrong.

2. Terms of the Lagrangian density and corresponding terms of the Hamiltonian density undergo different Lorentz transformations: Terms of the first case are Lorentz scalars, whereas terms of the second case are the T^{00} component of the energy-momentum tensor (see section 7.4, on p. 76).

   Conclusion: The tensorial form of the Pauli term should not be rejected as a candidate for the weak interactions in Lagrangian density.

3. A single configuration cannot correctly describe the state of an atom that has more than one electron (see section 7.5, on p. 78).

Conclusion: For a strange reason, the present community ignores evidence that was already known in the 1930s (see e.g. chapter XV of [68]). Furthermore, data on proton spin do not make a crisis for people who have studied this part of physics (see section 7.5, on p. 78).

4. The AB claims about the significance of the electromagnetic 4-potential and their topological arguments are wrong.

15.5 Electromagnetic Errors

Here is a list of erroneous elements of the present QED structure.

EMERR.1 Radiation fields and bound fields are different physical entities (see section 8.2 on p. 84).

EMERR.2 The electromagnetic 4-potential of radiation fields is not a 4-vector (see section 8.3 on p. 88).

EMERR.3 There is a relativistically consistent transformation of the radiation 4-potential. It is not its Lorentz transformation (see subsection 8.3.4 on p. 92.

EMERR.4 In VE, gauge transformations are forbidden. This means that a fundamental QED concept collapses (see section 8.4 on p. 93).

15.6 QCD Errors

Here is a list of experimental and theoretical QCD errors.

QCDERR.1 QCD has been constructed on an erroneous basis (see section 10.2 on p. 119).

QCDERR.2 The SM cannot explain the hard photon interaction with a nucleon (see subsection 10.5.1 on p. 129).

QCDERR.3 QCD cannot explain the nuclear force (see subsection 10.5.3 on p. 132).

QCDERR.4 QCD cannot explain the nuclear liquid drop structure (see subsection 10.5.5 on p. 135).

QCDERR.5 QCD cannot explain the nuclear tensor force (see subsection 10.5.6 on p. 136).

QCDERR.6 QCD cannot explain the EMC effect (see subsection 10.5.7 on p. 137).

QCDERR.7 QCD cannot explain why the binding energy of a strange quark in a baryon is stronger than its binding energy in the kaon. In contrast, the bond of the u, d quarks in the pion is stronger than their bond in the nucleon. (see subsection 10.5.8 on p. 138).

QCDERR.8 QCD in general and its asymptotic freedom in particular, cannot explain why the electron-proton total cross-section decreases with the increase of energy, whereas the high energy proton-proton total cross-section increases with energy (see subsection 10.5.11 on p. 142).

QCDERR.9 QCD in general and its asymptotic freedom in particular, cannot explain why the electron-proton elastic cross-section decreases very rapidly (and its relative portion becomes negligible) with the increase of energy, while the high energy proton-proton elastic cross-section increases with energy (see subsections 10.5.11 on p. 142 and 10.5.12, on p. 144).

QCDERR.10 QCD cannot explain why, at high energy, the π^+-proton elastic cross-section is smaller than the proton-proton elastic cross-section (see 10.5.13 on p. 145).

QCDERR.11 QCD cannot explain why antiquarks are pushed to the proton's peripheral region (see subsection 10.5.14 on p. 147).

QCDERR.12 QCD cannot explain the relatively large portion of $\bar{s}s$ pairs in the proton (see subsection 10.5.15 on p. 148).

QCDERR.13 QCD cannot explain why the proton has more $\bar{d}d$ pairs than $\bar{u}u$ pairs (see subsection 10.5.16 on p. 149).

QCDERR.14 QCD cannot explain the negative value of the neutron's mean square charge radius (see subsection 10.5.19 on p. 154).

QCDERR.15 QCD cannot explain the results of the polarized proton-proton scattering data. These data have been described as "the thorn in the side of QCD" (see subsection 10.5.20 on p. 155).

QCDERR.16 The QCD strong CP problem is unsettled (see subsection 10.5.21 on p. 156).

QCDERR.17 QCD supporters have no solid explanation for the failure to detect a genuine *strongly bound* pentaquark (see subsection 10.5.22 on p. 156).

Unfortunately, the community of QCD supporters has a quite simple "solution" to these intrinsic contradictions: Do not mention them in textbooks and reject (most) papers that mention these issues from mainstream journals.

15.7 Electroweak Errors

Here is a list of electroweak errors.

EWERR.1 The electroweak theory has been constructed on an erroneous basis (see section 11.2 on p. 172 to the end of section 11.6 on p. 183).

EWERR.2 SM textbooks still have not published an explicit form of the electroweak's Euler-Lagrange differential equations.

EWERR.3 A fortiori, SM textbooks still have not published solutions to the electroweak's Euler-Lagrange differential equations.

EWERR.4 The electroweak theory still has no coherent expression for the electromagnetic interaction of the $W^\pm$ particles (see subsection 11.6.2 on p. 184).

EWERR.5 The electroweak theory uses an erroneous quantum function for the Z particle (see subsection 11.6.5 on p. 189).

EWERR.6 The electroweak theory of the $W^\pm$ and Z particles cannot explain their pure leptonic decay (see subsection 11.6.6 on p. 189).

EWERR.7 The electroweak $(1 \pm \gamma^5)$ factor is unacceptable for a massive Dirac particle like the electron or the neutrino (see subsection 11.6.7 on p. 190).

EWERR.8 Because of the multitude of terms of the electroweak Lagrangian density, the Occam's razor principle casts very serious doubt on the acceptability of this theory (see subsection 11.6.8 on p. 192).

It is interesting to note that the above mentioned inherent problems of the electroweak second-order description of the W and Z particles are compatible with the Dirac lifelong objection to second-order quantum equations [130] .

15.8   Open Problems

This book does not aim to give the final word on any subject that it discusses. Physics has been developed by many people, and this issue will certainly remain valid in the future. In particular, new ideas and new concepts will arise as a result of new experimental evidence and the creativity and imagination of new people. The following list points out several topics that I regard as unfinished:

U.A Section 15.2 shows the general form of the Lagrangian density of a Dirac particle and its strong, electromagnetic, and weak interaction terms. Here the equations of motion take the form of the Dirac Hamiltonian, where each of the Lagrangian's interaction terms joins the equation with an opposite sign. It was pointed out in subsection 3.3.2 on p. 22 that one must accomplish further tasks to have a satisfactory theory:

   T.1 Satisfactory solutions to the equations of motion should be derived for specific systems.

   Considering hadrons, let us see why the task of T.1 is not as simple as it looks.

U.B The explicit detection of additional $\bar{q}q$ pairs in the proton is important experimental evidence. It proves that the quantum function that describes hadrons in general and nucleons in particular, should contain not only the valence quarks but also additional $\bar{q}q$ pairs of (at least) the three lightest u, d, s quarks. Hence, the proton structure should not be derived from the quantum function of three valence quarks (a three-body problem) but from the addition of functions that depend on quark-antiquark pairs:

$$\Psi = a_0\psi(u,u,d) + a_u\psi_u(u,u,d,u,\bar{u}) + a_d\psi_d(u,u,d,d,\bar{d})$$
$$+ a_s\psi_s(u,u,d,s,\bar{s}) + ..., \qquad (15.3)$$

where each a_k is a numerical coefficient and the ... symbol indicates functions of more than one additional $q\bar{q}$ pair. Mesons should also have additional $\bar{q}q$ pairs. This argument emphasizes the role of the Fock space.

Every quantum function on the right side of (15.3) should be written as a sum of terms, each of which depends on a specific quark configuration that is compatible with conservation laws, such as J^π and parity of the state (see section 7.5 on p. 78).

Another problem of a baryonic state calculation is the existence of the baryonic core and its closed shells of u and d quarks (see chapter 10). This core should be an element of a baryonic state calculation, and quantum functions of the u and d valence quarks should be orthogonal to functions of quarks of the closed shells.

For this reason, the explicit function of the core's closed shells of u and d quarks should also be calculated. These arguments explain why a genuine calculation of the baryonic structure is not a simple task.

U.C It is not clear whether and how the Bethe-Salpeter equation can be adapted to problems of more than two particles and/or to multiconfiguration states.

U.D A recalculation of the QED description of physical systems that uses a consistent Lagrangian density is required. For problems of the present QED Lagrangian density, see chapter 8 on p. 83.

U.E Electrodynamics has a tradition of accurate calculations of physical states and processes. The additional $\bar{q}q$ pairs in the proton's function (15.3) indicate that an accurate function of an electronic state should also comprise configurations of the actual electrons of the system and an addition of at least one electron-positron pair. The addition of this pair increases the number of acceptable configurations.

U.F Does the Darwin Lagrangian and the Breit interaction make an accurate replacement of bound fields? Consider a system of charged particles. Darwin constructed a mechanical-like Lagrangian for this system, where the electromagnetic interaction is replaced by the velocity-dependent expression

$$
L_{Darwin} = -\sum_j \sum_{i>j} \frac{e_j e_i}{R_{ij}} +
$$
$$
\sum_j \sum_{i>j} \frac{e_j e_i}{2R_{ij}} [\boldsymbol{v}_j \cdot \boldsymbol{v}_i + (\boldsymbol{v}_j \cdot \boldsymbol{n}_{ij})(\boldsymbol{v}_i \cdot \boldsymbol{n}_{ij})].
$$

$$(15.4)$$

Here $e_i, \boldsymbol{v}_i$ denote the charge and velocity of the ith particle, respectively, and i, j run on the particles' index. R_{ij} is the instantaneous distance between the ith and jth particles, and $\boldsymbol{n}_{ij}$ denotes the unit vector in the direction from the ith particle to the jth particle (see [3], pp. 179-182, [28], pp. 593-595). Like the standard structure of classical mechanics, the Darwin Lagrangian depends on the instantaneous values of the coordinates and the velocity of the charged particles *but is independent of the electromagnetic fields and their potentials.* In the quantum domain, the Darwin Lagrangian of a system of Dirac particles yields the Breit interaction (see [55], pp. 170, 195). Here, the Dirac velocity operator $\boldsymbol{\alpha}_i$ of each particle replaces the velocity $\boldsymbol{v}_i$.

Problem: Does the Breit interaction accurately describe the entire *2-body interaction of a quantum system?*

A comparison between experimental data of the He atom and results of calculations of states that account for the issues of item U.E can probably provide an answer to this problem.

U.G Section 11.8 on p. 195 explained why there are three kinds of weak interactions. These issues are derived from the three generations of leptons and quarks. The CKM matrix indicates analogous properties for the six quarks. The weak interaction Lagrangian density should correctly describe the generation dependence of weak interaction processes. Details of this problem still await an appropriate formulation.

U.H The main objective of this book is to analyze the veracity of field theories. For any given theory, the final word is the fit between solutions of its equations of motion and experimental data. However, physics is a quite mature science, and this book utilizes fundamental physical principles and their consequences. These issues are pointed out at the beginning of the book. Thus, numerical methods of solving the fields' equations of motion are generally outside the scope of this book.

Here, I wish to briefly discuss one numerical problem, which is useful for solving bound states of systems that comprise several spin-1/2 Dirac particles. Furthermore, this method provides a valuable insight into the correct structure of important quantum systems. Considering a bound state, the energy E is a good quantum number, and the time coordinate takes the form of the factor e^{-iEt}. Therefore, for finding out a solution for a quantum equation of a bound state, one can use the time-independent quantum functions of the Heisenberg picture and construct the Hilbert and Fock spaces.

Modern computers are based on principles of QM, and the computer epoch began several decades after the establishment of quantum theory. This evidence explains why physicists who worked in the early decades of quantum theory had no computers. For this reason, they strove to construct efficient mathematical tools for solving quantum problems. One of these tools is called angular momentum algebra, or Wigner-Racah algebra. This mathematical theory provides explicit formulas for integrals of angular coordinates. In particular, this mathematical theory reduces the number of free independent spatial coordinates by a factor of $1/3$. This reduction of the number of independent variables entails a dramatic simplification of the numerical problem and a large reduction of the computation time.

The simpler solution of the multi-configuration problem is an important property of the Wigner-Racah algebra. The significance of this achievement was discussed in section 7.5 on p. 78. In particular, the long-lasting problem called the *proton spin crisis* was automatically explained (see item 3. on p. 232). The immediate resolution of a long-lasting crisis proves the importance of the Wigner-Racah algebra and justifies its incorporation as a legitimate course of university physics.

U.I Readers may note that the *V-A* weak interaction terms (11.10) of this book denote the 3-vector γ^i and the axial 3-vector $\gamma^5\gamma^i$. In contrast, in this case, the SM uses 4-vector γ^μ and axial 4-vector $\gamma^5\gamma^\mu$ (see [67], pp. 308, 309). An examination of the data may decide which theory is correct.

U.J Finally, all other unsettled problems that I failed to mention above!

Chapter 16

Epilogue

This book primarily discusses physical topics. It shows new theoretical elements and points out many SM errors. The arguments rely on well-established experimental data and mathematical properties of physical principles. These are objective scientific issues. However, it is interesting to know the ideas of some SM supporters. A description of one example of this kind is described below. It was written as a stand-alone text, and I put its contents here.

16.1 Background

For more than 50 years, particle physics activity has run under the rule stipulating that it is strictly forbidden to discuss the possibility that there are errors in existing theories. The dictum "shut up and calculate" stems from this policy (readers may search the web for this "instruction"). This quasi-religious atmosphere has resulted in the present state, where the SM is full of errors. Hence, the kind of people who flourish in the present particle physics community learn things by heart without having a genuine understanding of the internal logic of their theories.

The last statement is quite harsh. Therefore, I wish to substantiate it with a description of simple cases. The primary element of the current field theory is the Lagrangian density. For example, an important mainstream textbook says: "All field theories used in current theories of elementary particles have Lagrangians of this form" (see [20], p. 300). I completely agree that this is the right course. The Lagrangian density of different interactions takes a different form.

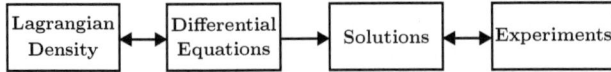

Figure 16.1: *Structure of a coherent field theory.*

Consider two important theories that describe the electromagnetic fields and the electron, called Maxwellian electrodynamics and the Dirac theory, respectively. These are relatively correct theories, and modern industry is based on them. Figure 16.1 illustrates the relations between crucial elements of each of these theories – a Lagrangian density, differential equations that are derived from the Lagrangian density, solutions of the differential equations, and relevant experimental results that fit the solutions.

All physicists and probably some mathematicians, chemists, and engineers have studied these theories. These people feel that they have studied solid, reliable theories.

Particle physicists study the electroweak theory, and they should recognize that the structure of this theory is completely different from that of Fig. 16.1 in the following ways:

F.1 Most textbooks do not show the full Lagrangian density of the electroweak theory.

F.2 No textbook shows the explicit form of the electroweak differential equations.

F.3 Clearly, no textbook shows solutions to these unknown equations.

F.4 Clearly, no textbook shows that the solutions to the unknown electroweak differential equations fit experimental data.

F.5 Apparently, members of the particle physics community are quite happy with this unfortunate plight. Indeed, many make declarations like the following: "The Standard Model: The most successful theory ever" [172, 173]. Another example is taken from a textbook: "Remarkably, the Standard Model provides a successful description of all current experimental data and represents one of the triumphs of modern physics" (see [79], p.1). Furthermore, the Wikipedia policy represents the current consensus. As of August 2021, this grave situation of the electroweak theory is not mentioned on the Wikipedia list of unsolved problems in physics [93].

These facts substantiate my assertion that particle physicists study their theories in a quasi-religious form, and they do not strive to find logical coherence between elements of their theories. The events that are described below provide another example that substantiates my claim.

16.2   Meeting a Particle Physics Expert

Recently, I have exchanged several emails with a particle physics expert, referred to as X here. I do not mention his name because I wish to remove all personal aspects from the discussion. A reputable source says that X is regarded as a brilliant and rising young particle physicist, and he has been awarded many prizes.

16.2.1   Problem A

I sent X this problem: "The electroweak theory regards the $W^\pm$ particles as elementary charge-carrying particles. Can you please explain why unlike the Dirac electron theory, the electroweak theory violates Maxwellian electrodynamics because it is unable to provide an explicit expression for the $W^\pm$ 4-current that satisfies the continuity equation?" (The full email is in [174].) Readers should note that the particle's density is the 0-component of its 4-current.

X has found a simple answer : There is no problem with the $W^\pm$ particles because they are unstable! (See his email [175].)

I was surprised at his answer, and I am quite sure that it is his invention and cannot be found in textbooks. I have told him that the muon negates his assertion: "The muon and the W have some common properties: They are electrically charged particles; they are unstable; they decay due to weak interactions." The Dirac theory provides a coherent expression for the muon density. Furthermore, "people who have built muon accelerators" used its density "for the beam construction and the particle's acceleration." Hence, why should the electroweak theory not do the same for the $W^\pm$? (The full text is in [176].) These arguments have not convinced X, and he restates to point below.

> "The muon current you wrote is not conserved and
> neither is any W boson current that one can write
> (whether it is the one that couples to electromag-
> netism or any other improved current). Therefore I do
> not understand what is the issue that you are bring-
> ing up. BTW this material is completely standard
> since the early 70s."

I wish to add two fundamental arguments that support my opinion, which are as follows:

A.1 Let us examine the inhomogeneous Maxwell equations (see
[3], p. 79; [28], p. 551)

$$F^{\mu\nu}_{,\nu} = -4\pi j^{\mu}. \tag{16.1}$$

Taking the 4-divergence of (16.1), one uses the antisymme-
try of the tensor of the electromagnetic fields and proves the
continuity equation of the charge-carrying particle

$$0 = F^{\mu\nu}_{,\nu,\mu} = -4\pi j^{\mu}_{,\mu}. \tag{16.2}$$

These arguments prove that Maxwell equations of the electromagnetic fields *impose* the continuity equation on *every* theory of an elementary charge-carrying particle (see also [20],
p. 341). Hence, the conclusion below can be made.

> Every theory of a charge-carrying elementary particle
> must satisfy the continuity equation. A violation of
> the continuity equation means a violation of Maxwell
> equations.

The continuity equation is studied in an undergraduate
course on electrodynamics.   The muon carries an electric
charge. Hence, its theory must prove that it satisfies the continuity equation (16.2). The Dirac equation of the muon is
okay.

A.2 Let us turn to the muon experimental aspects. This information is included in an elementary particle undergraduate
course. Figure 16.2 illustrates the tracks seen in the photographic emulsion of the first experimental muon detection
(see [23], p. 4). The full muon track is seen until it decays. Its decay is an instantaneous process, and the track of

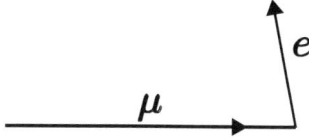

Figure 16.2: *Illustration of muon decay.*

the outgoing electron makes an angle with the muon's track. (Each of the two outgoing neutrinos is a chargeless particle that is unseen in the photograph.) *These data prove that before decaying, the muon behaves like an ordinary charged Dirac particle.* Its charge is the same as that of the electron and it ionizes atoms of the photographic emulsion. Hence, it satisfies Maxwellian electrodynamics. This means that it satisfies the continuity equation (16.2). As stated above, the Dirac equation of the muon is okay.

My Conclusion: A brilliant and arising young particle physicist who has been awarded many prizes simply does not know important physical properties of Maxwell equations and the muon. For example, he does not know that the Dirac equation for the muon provides an expression for the 4-current that satisfies the continuity equation. He is also unaware of basic information of experimental particle physics.

16.2.2   Problem B

Another excerpt from what I have sent to X refers to the electromagnetic interaction term of the Lagrangian density of the electroweak theory of the $W^\pm$. I have told him that the quadratic factor of the 4-potential of an electromagnetic interaction term like

$$\mathcal{L}_{int} = a\phi^\dagger \chi A A \qquad (16.3)$$

violate Maxwellian electrodynamics. Variables of (16.3) are as follows: a is a numerical constant, ϕ, χ denote quantum functions of an electrically charged particle, A is the electromagnetic 4-potential, and the indices that show how this expression is contracted and produces a Lorentz scalar are suppressed. The electromagnetic interaction term of the electroweak theory of $W^\pm$ takes the form of (16.3), where ϕ, χ are entries of the W 4-vector.

The reason for my assertion is quite simple: The Euler-Lagrange equation of the rth quantum function is

$$\frac{\partial \mathcal{L}}{\partial \psi_r} - \frac{\partial}{\partial x^\mu} \frac{\partial \mathcal{L}}{\partial (\partial \psi_r / \partial x^\mu)} = 0, \qquad (16.4)$$

(see e.g. [20], p. 300; [26], p. 15). Here, ψ_r denotes the generalized coordinates of the Lagrangian density. In electrodynamics, the 4-potential A_μ is regarded as the generalized coordinate of the Lagrangian density (see [3], p. 78; [28], p. 596). Hence, an application of the derivative operator of the first term of the Euler-Lagrange equation (16.4) to the interaction term of (16.3) yields equations of the electromagnetic fields that explicitly depend on the 4-potential A_μ. This is a sheer violation of gauge invariance. X has responded with contradictory declarations.

> "There are quadratic pieces in the Standard Model for similar reasons. This does not contradict anything and it does not contradict the gauge invariance of Maxwell theory."

He emphasizes his argument.

> "Lagrangians containing quadratic fields in A_μ are pervasive and we know why they are correct. In the theory of superconductivity this is the basic thing you learn in the first course on the topic:
>
> $$L = |\partial \Phi - A_\mu \Phi|^2 + V(|\Phi|). \qquad (16.5)$$
>
> Please check the first equation on the Wikipedia page for superconductivity
> `https://en.wikipedia.org/wiki/`
> `GinzburgLandau_theory`
> Expanding it out you see it has a quadratic piece in A. It contradicts nothing – it is as established part of physics as classical mechanics.
> There are quadratic pieces in the Standard Model for similar reasons. This does not contradict anything and it does not contradict the gauge invariance of Maxwell theory. I teach it to graduate students in *University Y* and this has been standard material for at least 100 years." (Remark: University Y belongs to the top 10 in the ranking list of US universities.)

A straightforward examination of the quotation from X's arguments proves the following things:

P.1 The Lagrangian density is the primary QFT quantity. It turns out that X does not understand the significance of the restriction Req.XI on p. 227, which forbids the factor e^2 of the charge, and Req.XII, which forbids the quadratic factor of the 4-potential A_μ.

P.2 I use an elementary derivative operation and prove an inherent error of the electroweak theory.

P.3 X demonstrates his quasi-religious style and quotes from his "holy scriptures." He argues that there is no problem with the SM just because the same error exists in many places...

> **My Conclusion:** A brilliant and arising young physicist who has been awarded many prizes, ignores a straightforward solid argument that uses the Euler-Lagrange equation and fundamental laws of math. He is absolutely sure that he is right just because the same error exists in other SM places. (See also the discussion about the DAEI in subsections 11.6.3, p. 186 and 11.6.4 on p. 188.)

16.2.3 Problem C

I have asked X's opinion on the proof that QCD has been constructed on an erroneous basis and sent him [177]. He did not answer this problem and terminated the correspondence.

> **My Conclusion:** A brilliant and arising young physicist who has been awarded many prizes cannot refute a two-page proof showing the erroneous basis of QCD.

16.3 Conclusions

Here is my opinion of the previous issues.

> A brilliant and arising young physicist who has been
> awarded many prizes is extremely eager to prove the
> error-free structure of the SM. In doing so, he makes
> gross errors. This is another example that illustrates
> my claim about the quasi-religious behavior of mem-
> bers of the particle physics community. They never
> admit that there is a flaw in their theories, and they
> do not care about the logical coherence of their theo-
> ries; they study the SM by heart.

Similar remarks on the pernicious quasi-religious aspects of the
particle physicists activity have been published by another person,
who says that he is a qualified theoretical physicist [178].

Index

        81, 113–115, 117–127,
        129, 132, 134, 137,
        138, 150, 153, 156–
        160, 162–164, 167,
        169, 171, 178, 179,
        195–197, 205, 206,
        212, 213, 215, 220,
        225, 229–231, 234, 236
Supersymmetry, *see* SUSY
SUSY, 5, 211, 212

Tensor, *see* 4-tensor
Tensor force, 136, 137, 233
Tensor interaction, 177, 178

Ultrarelativistic, 180, 198, 199

Van der Waals, iii, 133
Vector
     axial, 58, 136, 175, 179
     meson, 130
     polar, 58, 137
     potential, 88, 106
Velocity
     group, 48, 49
     operator, 62, 100, 238
     phase, 48, 49

W (particle), 63, 65, 68, 69,
        171, 175, 184–190,
        194, 203, 212, 219–
        223, 228, 230, 235,
        243, 245
Weak interactions, xiv, 4, 5, 7,
        9, 22, 35, 39, 65, 69,
        71, 113–115, 164, 166,
        171–184, 191, 195–
        197, 199, 200, 212,
        213, 215, 217, 220–
        222, 229–232, 236,
        238, 239, 243

Yukawa, 133

Z (particle), 45, 63, 203, 212,
        221, 223, 228, 235

Bibliography

[1] Wigner, E. (1960) The unreasonable effectiveness of mathematics in the natural sciences. Comm. Pure Appl. Math., 13, 1-14.

[2] J. D. Bjorken and S. D. Drell, *Relativistic Quantum Mechanics* (McGraw-Hill, New York, 1964).

[3] L. D. Landau and E. M. Lifshitz, *The Classical Theory of Fields* (Elsevier, Amsterdam, 2005).

[4] P. A. M. Dirac, Phys. Rev. **74**, 817 (1948).

[5] https://en.wikipedia.org/wiki/Mathematics

[6] https://plato.stanford.edu/entries/models-science/#ModeTheo

[7] https://en.wikipedia.org/wiki/Standard_Model.

[8] S. S. M. Wong, *Introductory Nuclear Physics* (Wiley, New York, 1998).

[9] A. de-Shalit and I. Talmi, *Nuclear Shell Theory* (Academic Press, New York, 1963).

[10] J. Janecke, Atomic Data and Nuclear Data Tables, **17**, 455 (1976).

[11] E. Comay and I Kelson, Atomic Data and Nuclear Data Tables, **17**, 463 (1976).

[12] J. Janecke and B. P. Eynon, Atomic Data and Nuclear Data Tables, **17**, 467 (1976).

[13] https://en.wikipedia.org/wiki/Occam\%27s_razor

[14] M. E. Peskin and D. V. Schroeder, *An Introduction to Quantum Field Theory* (Addison-Wesley, Reading Mass, 1995).

[15] E. Merzbacher, Quantum Mechanics, second edition (John Wiley, New York, 1970).

[16] Wigner, E. (1939). On Unitary Representations of the Inhomogeneous Lorentz Group. Ann. Math. 40, 149-204.

[17] Rohrlich, F. (2007). Classical charged particle. (New Jersey: World Scientific).

255

[18] L. I. Schiff, *Quantum Mechanics* (McGraw-Hill, New York, 1955).

[19] Einstein, A (2000). Albert Einstein in his own words (New York: Portland House).

[20] S. Weinberg, *The Quantum Theory of Fields,* Vol. I (Cambridge University Press, Cambridge, 1995).

[21] Landau, L. D., & Lifshitz, M. E. Mechanics (Pergamon, Oxford, 1960)

[22] H. Goldstein, C. Poole & J. Safko, Classical Mechanics, 3rd edition (Addison Wesley, San Francisco, 2002).

[23] D. H. Perkins, *Introduction to High Energy Physics*, Menlo Park CA, Addison-Wesley, 1987.

[24] P. A. M. Dirac, *The Principles of Quantum Mechanics*, (Oxford University Press, London, 1958).

[25] https://en.wikipedia.org/wiki/Double-slit_experiment#Interference_of_individual_particles

[26] J. D. Bjorken and S. D. Drell, *Relativistic Quantum Fields* (McGraw-Hill, New York; 1965).

[27] Landau, L. D., & Lifshitz, M. E. Quantum Mechanics (Pergamon, London, 1959)

[28] Jackson, J. D. (1975). Classical Electrodynamics. New York: John Wiley.

[29] P.A. Zyla et al. (Particle Data Group), Prog. Theor. Exp. Phys. 2020, 083C01 (2020).

[30] L. C. Tu and J. Luo, Metrologia **41** S136 (2004) https://www.researchgate.net/publication/228689980

[31] E. M. Purcell and D. J. Morin, *Electricity and Magnetism* (Cambridge University Press, Cambridge, 2013).

[32] P. Goddard and D. I. Olive, Rep. Prog. Phys., 41, 1357 (1978).

[33] P. A. M. Dirac, Proc. Royal Soc. **A133**, 60 (1931).

[34] D. Milstead and E.J.Weinberg https://pdg.lbl.gov/2017/reviews/rpp2017-rev-mag-monopole-searches.pdf

[35] E. Comay *Lett. Nuovo Cimento* **43**, 150 (1985).

[36] E. Comay, Nuovo Cimento, **80B**, 159 (1984).

[37] E. Comay, Nuovo Cimento, **110B**, 1347 (1995).

[38] F. Halzen and A. D. Martin, Quarks and Leptons, An Introductory Course in Modern Particle Physics (John Wiley, New York,1984)

[39] T. H. Bauer, R. D. Spital, D. R. Yennie, and F. M. Pipkin, Rev. Mod. Phys. **50** 261 (1978).

[40] E. Comay, Elect. J. Theor. Phys., **9**, 93-118 (2012).
http://www.ejtp.com/articles/ejtpv9i26p93.pdf

[41] E. Comay, published in *Has the Last Word Been Said on Classical Electrodynamics?* Editors: A Chubykalo, V Onoochin, A Espinoza, and R Smirnov-Rueda (Rinton Press, Paramus, NJ, 2004). (The Article's title is "A Regular Theory of Magnetic Monopoles and Its Implications".)

[42] O. Comay, Science or Fiction? The Phony Side of Particle Physics (S. Wachtman's Sons, CA, 2014).

[43] Comay, E. (2020). A Critical Study of Quantum Chromodynamics and the Regular Charge-Monopole Theory. Physical Science International Journal, 24(9), 18-27.
https://doi.org/10.9734/psij/2020/v24i930213

[44] P. A. M. Dirac, Proc. Roy. Soc. Lond., **A117**, 610 (1928).

[45] https://en.wikipedia.org/wiki/Dirac_equation

[46] S. Weinberg, *The Quantum Theory of Fields,* Vol. II (Cambridge University Press, Cambridge, 1996).

[47] G. Sterman, *An Introduction to Quantum Field Theory* (Cambridge University Press, Cambridge, 1993). (See p. 518).

[48] W. N. Cottingham and D. A. Greenwood, "An Introduction to the Standard Model of Particle Physics" (Cambridge University Press, Cambridge, 2007). Second Edition).

[49] https://en.wikipedia.org/wiki/Electroweak_interaction#After_electroweak_symmetry_breaking

[50] Sternberg, S. (1994). Group Theory and Physics. Cambridge: Cambridge University Press. (See pp. 143-150.)

[51] Schweber S. S. (1964). An Introduction to Relativistic Quantum Field Theory. New York: Harper & Row. (See pp. 44-53.)

[52] Pauli W. (1941). Relativistic Field Theories of Elementary Particles. Rev. Mod. Phys. 13, 203-232.

[53] V. B. Berestetskii, E. M. Lifshitz and L. P. Pitaevskii, *Quantum Electrodynamics* (Pergamon, Oxford, 1982).

[54] H. Dehmelt, Physica Scripta, **T22**, 102 (1988).

[55] H. A. Bethe and E. E. Salpeter *Quantum Mechanics of One- and Two-Electron Atoms* (Springer, Berlin, 1957).

[56] A. Messiah, *Quantum Mechanics*, V. 1 (North Holland, Amsterdam, 1967).

[57] https://en.wikipedia.org/wiki/Wave%E2%80%93particle_duality

[58] C. A. Coulson, Waves (Oliver and Boyd, Edinburgh, 1961).

[59] https://en.wikipedia.org/wiki/Double-slit_experiment

[60] https://en.wikipedia.org/wiki/Gram%E2%80%93Schmidt_process

[61] https://arxiv.org/abs/1203.4051

[62] https://en.wikipedia.org/wiki/Fock_space

[63] W. Shockley and R. P. James, Phys. Rev. Lett., **18**, 876 (1967).

[64] S. Coleman and J. H. Van Vleck, Phys. Rev., **171**, 1370 (1968).

[65] E. Comay, Am. J. Phys., **64**, 1028 (1996).

[66] R. Adler, M. Bazin and M. Schiffer, Introduction to General Relativity (McGraw-Hill, New York, 1965).

[67] Griffiths D. Introduction to elementary particles. 2nd edition. Wiley-VCH, Weinheim; 2008.

[68] E. U. Condon and G. H. Shortley, *The Theory of Atomic Spectra* (University Press, Cambridge, 1964).

[69] J. C. Slater, Quantum Theory of Atomic Structure, Vol. II (McGraw-Hill, New York, 1960).

[70] G. R. Taylor and R. G. Parr, Proc. Natl. Acad. Sci. **38**, 154 (1952).

[71] C. F. Bunge, Phys. Rev. A **14**, 1965 (1976).

[72] L. C. Biedenharn, J. D. Louck, *The Racah-Wigner Algebra in Quantum Theory* (Cambridge University Press, Cambridge, 1984).

[73] A. Messiah, *Quantum Mechanics*, V. 2 (North Holland, Amsterdam, 1964).

[74] The Wikipedia configuration item
https://en.wikipedia.org/wiki/Electron_configuration

[75] The Wikipedia item on the *Proton Spin Crisis.*
https://en.wikipedia.org/wiki/Proton_spin_crisis

[76] P. A. M. Dirac, Scientific American, **208**, 45, May 1963.

[77] R. P. Feynman, *QED, The Strange Theory of Light and Matter* (Penguin, London, 1990).

[78] L. H. Ryder, *Quantum Field Theory* (Cambridge University Press, Cambridge, 1997).

[79] M. Thomson, *Modern Particle Physics* (Cambridge University Press, Cambridge, 2013).

[80] R. P. Feynman, R. B. Leighton and M. Sands, *The Feynman Lectures on Physics*, V. II (Addison-Wesley, Reading Mass., 1965).

[81] E. Comay, *Acta Physica Polonica A*, **133**, 1294 (2018).
http://przyrbwn.icm.edu.pl/APP/PDF/133/app133z5p26.pdf.

[82] Y. Aharonov and D. Bohm Phys. Rev., **115**, 485 (1959).

[83] Y. Aharonov and D. Bohm, Phys. Rev., **123**, 1511 (1961).

[84] https://en.wikipedia.org/wiki/Aharonov%E2%80%93Bohm_effect

[85] E. Comay, Physical Science International Journal, **25**, 52 (2021).
      https://www.journalpsij.com/index.php/PSIJ/article/view/30238/
      56734

[86] A. Tonomura, N. Osakabe, T. Matsuda, T. Kawasaki, J. Endo, S.
      Yano, and H. Yamada, *Phys. Rev. Lett.* **56**, 792 (1986).

[87] https://en.wikipedia.org/wiki/Quantum_entanglement

[88] A. C. T. WU and C. N. Yang, Int. J. Mod. Phys. A **21**, 3235 (2006).

[89] H. Fritzsch, M. Gell-Mann and H. Leutwyler, Phys. Lett. **47B**, 365
      (1973).

[90] H. Fritzsch, CERN Courier, Sep 27, 2012.
      http://cerncourier.com/cws/article/cern/50796

[91] C. W. Patterson, Phys. Rev. A **100**, 062128 (2019).

[92] S. Wolfram, "Remembering Murray Gell-Mann"
      https://writings.stephenwolfram.com/2019/05/remembering-
      murray-gell-mann-1929-2019-inventor-of-quarks/

[93] https://en.wikipedia.org/wiki/List_of_unsolved_problems_in_
      physics#High-energy_physics/particle_physics.

[94] E. M. Henley and A. Garcia, *Subatomic Physics* (World Scientific, New
      Jersey, 2007)

[95] H. Haken and H. C. Wolf, *Molecular Physics and Elements of Quantum Chemistry* 2nd edition (Springer, Berlin, 2004).

[96] F. Wilczek, Nature, **445**, 156 (2007).

[97] N. Ishii, Aoki and T. Hatsuda, Phys. Rev. Lett., **99**, 022001 (2007).

[98] J. J. Aubert et al. (EMC), Phys. Lett. **123B**, 275 (1983).

[99] A. Bodek et al., Phys. Rev. Lett. **50**, 1431 (1983).

[100] J. B. Pendry, J. Phys. **C13**, 3357, (1980).

[101] J. Arrington et al., J. Phys. Conference Series, **69**, 012024 (2007).

[102] Particle Data Group (2012)
       http://pdg.lbl.gov/2012/reviews/rpp2012-rev-cross-section-
       plots.pdf

[103] G. Antchev20 et al., TOTEM Collaboration, Eur. Phys. J. C **79**, 103
       (2019).
       https://doi.org/10.1140/epjc/s10052-019-6567-0

[104] M. Alberg, Prog. Part. Nucl. Phys. **61** 140 (2008).

[105] P. E. Reimer and the Fermilab SeaQuest Collaboration, EPJ Web of Conferences **113**, 05012 (2016)
https://www.epj-conferences.org/articles/epjconf/pdf/2016/08/epjconf_fb2016_05012.pdf

[106] P. E. Reimer,
https://www.anl.gov/event/measurement-of-the-flavor-asymmetry-in-the-protons-sea-quarks

[107] W. Scheinast et al., HZDR Institute.
https://www.hzdr.de/FWK/jb02/PDF/page8.pdf

[108] W. Scheinast et al., Phys. Rev. Lett. **96**, 072301 (2006).

[109] The Wikipedia item on the *proton spin crisis*
https://en.wikipedia.org/wiki/Proton_spin_crisis

[110] E. Comay, Prog. in Phys. 1, 75 (2011).

[111] A. D. Krisch, Hard collisions of spinning protons: Past, present and future, The European Physical Journal A**31**, 417-423 (2007).

[112] See the Wikipedia item
https://en.wikipedia.org/wiki/Strong_CP_problem

[113] https://en.wikiquote.org/wiki/Paul_Dirac

[114] C. Gignoux, B. Silvestre-Brac and J. M. Richard, Phys. Lett. **193**, 323 (1987).

[115] H. J. Lipkin, Phys. Lett. **195**, 484 (1987).

[116] https://cerncourier.com/a/four-labs-find-five-quark-particle/

[117] N. Akopov et al. (HERMES Collaboration) Phys. Rev. **D91**, 057101 (2015). A preprint at https://arxiv.org/pdf/1412.7317.pdf.

[118] R. Aaij et al. (LHCb Collaboration), Phys. Rev. Lett. **117**, 082002 (2016).

[119] https://arxiv.org/abs/1507.03414

[120] https://home.cern/news/news/physics/lhcb-experiment-discovers-new-pentaquark

[121] https://en.wikipedia.org/wiki/Pentaquark

[122] https://en.wikipedia.org/wiki/Noble_gas

[123] The Ω^- discovery.
http://www.hep.fsu.edu/~wahl/satmorn/history/Omega-minus.asp.htm

[124] E. Witten, Phys. Rev. **D30**, 272 (1984).

[125] K. Han, et al., Phys. Rev. Lett. **103**, 092302 (2009).

[126] The Wikipedia QCD item.
https://en.wikipedia.org/wiki/Quantum_chromodynamics

[127] https://en.wikipedia.org/wiki/Three-jet_event

[128] https://home.cern/news/news/physics/four-decades-gluons

[129] W. Rindler, *Special Relativity*, (Oliver and Boyd, Edinburgh, 1960).

[130] P. A. M. Dirac, *Mathematical Foundations of Quantum Theory*, Ed. A. R. Marlow (Academic, New York, 1978).

[131] R. P. Feynman and M. Gell-Mann, Phys. Rev. **109**, 193 (1958).

[132] G. Wentzel, Quantum Theory of Fields (Interscience, New York, 1949)

[133] Comay E. (2016). A Theory of Weak Interaction Dynamics. Open Access Library Journal, 3, 1-10.
https://www.scirp.org/journal/PaperInformation.aspx?paperID=72788

[134] Comay E. (2017). Further Aspects of Weak Interaction Dynamics. Open Access Library Journal. 4 1-11.
https://www.scirp.org/journal/PaperInformation.aspx?PaperID=74373

[135] Comay E. (2019). Differences Between Two Weak Interaction Theories. Phys. Sci. Int. J. 21(1), 1.
http://www.journalpsij.com/index.php/PSIJ/article/view/30091/56456

[136] C. G. Darwin, Proc. Roy. Soc. Lond. **A118**, 654 (1928).

[137] K. Hagiwara, R.D. Peccei, D. Zeppenfeld and K. Hikaso, Nuc. Phys. **B282**, 253 (1987).

[138] K. Hagiwara, J. Woodside, D. Zeppenfeld, Phys. Rev. **D41**, 2113 (1990).

[139] V. M. Abazov et al. (D0 collaboration), Phys. Lett. **B718**, 451 (2012)

[140] G. Aad et al. (ATLAS Collaboration), Phys. Lett. **B712**, 289 (2012).

[141] W. Pauli and V. Weisskopf, Helv. Phys. Acta, **7**, 709 (1934). English translation: A. I. Miller *Early Quantum Electrodynamics*, Cambridge, University Press, 1994. pp. 188-205.

[142] Formaggio JA, ZellerGP. From eV to EeV: Neutrino cross-sections across energy scales. Rev. Mod. Phys. 2012;84:1307-1341.

[143] https://en.wikipedia.org/wiki/Occam%27s_razor

[144] A. Salam, Nobel Lecture
https://www.nobelprize.org/uploads/2018/06/salam-lecture.pdf

[145] S. M. Bilenky, Phys. Part. Nuclei **46**, 475 (2015).

[146] M. Srednicki *Quantum Field Theory* (Cambridge University Press, Cambridge, 2007).

[147] Tanabashi M. et al. (Particle Data Group), Review of Particle Physics. Phys. Rev. D. 2018;98(030001):1-1989. (See p. 10 of http://pdg.lbl.gov/2019/reviews/rpp2019-rev-cross-section-plots.pdf)

[148] A Wikipedia item. Available: https://en.wikipedia.org/wiki/Black_hole

[149] A Wikipedia item. Available: https://en.wikipedia.org/wiki/Quasar

[150] A Wikipedia item. Available: https://en.wikipedia.org/wiki/Blazar

[151] S. Mertens, Direct Neutrino Mass Experiments. Journal of Physics: Conference Series, 2016;718(022013):1-9. Available: https://iopscience.iop.org/article/10.1088/1742-6596/718/2/022013/pdf

[152] Smponias T, Kosmas OT. High Energy Neutrino Emission from Astrophysical Jets in the Galaxy. Advances in High Energy Physics. 2015:Article ID 921757;1-7. Available: http://downloads.hindawi.com/journals/ahep/2015/921757.pdf

[153] Smponias T, Kosmas O. Neutrino Emission from Magnetized Microquasar Jets. Advances in High Energy Physics. 2017:Article ID 4962741;1-7. Available: http://downloads.hindawi.com/journals/ahep/2017/4962741.pdf

[154] The IceCube Collaboration. IceCube neutrinos point to long-sought cosmic ray accelerator. 2018. Available: https://icecube.wisc.edu/news/view/586

[155] Bergstrom L., Non-baryonic dark matter: observational evidence and detection methods. Rep. Prog. Phys. 2000;63:793-841. Available: https://iopscience.iop.org/article/10.1088/0034-4885/63/5/2r3/pdf

[156] Overduin JM, Wesson PS. Dark matter and background light. Phys. Rep. 2004;402:267-406. Available: https://www.sciencedirect.com/journal/physics-reports/vol/402/issue/5

[157] A Wikipedia item. Available: https://en.wikipedia.org/wiki/Dark_matter

[158] Di Valentino E, Melchiorri A, Silk J. Planck evidence for a closed Universe and a possible crisis for cosmology. Nature Astronomy. 2019;4:196-203. Available: https://www.nature.com/articles/s41550-019-0906-9.pdf

[159] Luminet J.-P, Weeks J, Riazuelo A, Lehoucq R, Uzan J.-P. Dodecahedral space topology as an explanation for weak wide-angle temperature correlations in the cosmic microwave background. Nature. 2003;425:593-595. Available: https://arxiv.org/abs/astro-ph/0310253

[160] A NASA publication. Available: https://map.gsfc.nasa.gov/universe/uni_shape.html

[161] Vardanyan M, Trotta R, Silk J. How flat can you get? A model comparison perspective on the curvature of the Universe. Monthly Notices of the Royal Astronomical Society. 2009;397(1):31-444. Available: https://academic.oup.com/mnras/article/397/1/431/1006985

[162] P. B. Pal, *Am. J. Phys.*, **79**, 485 (2011).

[163] The Wikipedia Neutrinoless Double Beta decay Item. (2021). https://en.wikipedia.org/wiki/Neutrinoless_double_beta_decay

[164] The Wikipedia item on the double neutrino β decay https://en.wikipedia.org/wiki/Double_beta_decay

[165] M. Agostini et al. (GERDA Collaboration), Phys. Rev. Lett. bf 125, 252502 (2020).

[166] The SUSY Wikipedia item of January 2021. https://en.wikipedia.org/wiki/Supersymmetry

[167] The GUT Wikipedia item of August 2021. https://en.wikipedia.org/wiki/Grand_Unified_Theory

[168] https://en.wikipedia.org/wiki/Proton_decay

[169] G. Goldhaber and J. E. Wiss, Ann. Rev. Nucl. Part. Sci. **30**, 337 (1980).

[170] https://en.wikipedia.org/wiki/Standard_Model

[171] https://www.lhc-closer.es/taking_a_closer_look_at_lhc/0.higgs_particle

[172] http://news.fnal.gov/2011/11/the-standard-model-the-most-successful-theory-ever/

[173] http://www.elliottmccrory.com/wp/2016/the-standard-model-the-most-successful-theory-ever/

[174] http://www.tau.ac.il/~elicomay/X_WPM1.pdf

[175] http://www.tau.ac.il/~elicomay/X_210312.txt

[176] http://www.tau.ac.il/~elicomay/X_From_EC_02.pdf

[177] https://www.tau.ac.il/~elicomay/dpp_new.pdf

[178] https://www.amazon.com/review/R1ZJZAZXO7G9QA

Made in the USA
Las Vegas, NV
06 January 2022

40545952R20155